AF713408

Internally Compensated LDO Regulators for Modern System-on-Chip Design

José María Hinojo · Clara Luján Martínez
Antonio Torralba

Internally Compensated LDO Regulators for Modern System-on-Chip Design

Springer

José María Hinojo
Electronic Engineering Department
Universidad de Sevilla
Seville
Spain

Antonio Torralba
Electronic Engineering Department
Universidad de Sevilla
Seville
Spain

Clara Luján Martínez
Electronic Engineering Department
Universidad de Sevilla
Seville
Spain

ISBN 978-3-319-75410-9 ISBN 978-3-319-75411-6 (eBook)
https://doi.org/10.1007/978-3-319-75411-6

Library of Congress Control Number: 2018933014

© Springer International Publishing AG, part of Springer Nature 2018
This work is subject to copyright. All rights are reserved by the Publisher, whether the whole or part of the material is concerned, specifically the rights of translation, reprinting, reuse of illustrations, recitation, broadcasting, reproduction on microfilms or in any other physical way, and transmission or information storage and retrieval, electronic adaptation, computer software, or by similar or dissimilar methodology now known or hereafter developed.
The use of general descriptive names, registered names, trademarks, service marks, etc. in this publication does not imply, even in the absence of a specific statement, that such names are exempt from the relevant protective laws and regulations and therefore free for general use.
The publisher, the authors and the editors are safe to assume that the advice and information in this book are believed to be true and accurate at the date of publication. Neither the publisher nor the authors or the editors give a warranty, express or implied, with respect to the material contained herein or for any errors or omissions that may have been made. The publisher remains neutral with regard to jurisdictional claims in published maps and institutional affiliations.

Printed on acid-free paper

This Springer imprint is published by the registered company Springer International Publishing AG part of Springer Nature
The registered company address is: Gewerbestrasse 11, 6330 Cham, Switzerland

To our families

Preface

Linear regulators, and more concretely, low-dropout (LDO) linear regulators, are essential blocks in power management systems, as they are able to generate precise and stable, low-noise, supply voltages, making them the right choice for supplying sensitive blocks in analogue, mixed-signal and radio-frequency circuits. In addition, LDO regulators can be completely integrated with no need of external components, what translates into significant savings in area and cost. Therefore, LDO regulators are attracting the attention from the scientific community, despite their use has some drawbacks. To mention, the maximum theoretical efficiency of LDO regulators is lower than that of switched-capacitor or inductor-based solutions. Even more, using internal compensation, the system's dominant pole is located at an internal node, and the first non-dominant pole (located at the output node) depends on the load. This introduces stability concerns, as load variations result on a frequency displacement of the first non-dominant pole that degrades system stability.

This book tackles the study of low-power, internally compensated LDO (IC-LDO) regulators. Through the book chapters, the reader will find an extensive revision of circuits and techniques proposed in the literature to deal with stability and transient response, especially those compatible with a low power consumption. In addition, the authors propose some innovative circuits and techniques, which are analysed, tested and compared with the state of the art.

Chapter 1 introduces the challenges that power management and distribution face in modern System-on-Chips (SoCs). It shows how linear voltage regulators, and specifically LDO regulators, have become critical blocks.

Chapter 2 introduces the IC-LDO regulator, starting with the classical topology and analysing how it is able to regulate the output voltage. In addition, it gives an overview of the main issues that a designer must take into account when undertaking its design. In every case, a thorough review of the techniques proposed in the literature is done. This chapter ends with a comparison of recently published IC-LDO regulators. The results of this comparison allow the reader to acquire confidence with typical values for each of the design parameters, and to understand the main trade-offs of the design process.

Next, in Chap. 3, a new technique is proposed to improve the stability of IC-LDO regulators. It is based on the classical Miller compensation scheme and takes advantage of an adaptive nulling resistor to extend the stability of Miller-compensated IC-LDO regulators to a wide range of load currents. This technique has been implemented in an IC-LDO regulator, and its measured performances are compared to other regulators recently published.

Chapter 4 presents an ultra-low-power IC-LDO regulator based on the classical topology, where a buffer has been added to drive the large parasitic capacitance of the pass transistor. In order to reduce the quiescent power consumption as much as possible, this buffer has been implemented with a class AB voltage follower. Just like in previous chapters, measurements of a regulator implementing the proposed contribution are shown, and its performances are compared to the state of the art.

In Chap. 5, the Flipped Voltage Follower (FVF) cell is presented as an alternative topology to implement an IC-LDO regulator. This cell and its cascode version (Cascode Flipped Voltage Follower or CAFVF) are analysed, and their behaviour as LDO regulator is compared to that of the classical topology. The FVF cell and its related circuits have been widely used to build LDO regulators. Some improvements to the basic cell have been proposed in the literature to enhance its stability and transient performances. Once again, a comparison of recently published papers that use the FVF or derived cells is presented. At the end of Chap. 5, another contribution of the authors is presented that improves the transient response of FVF-based LDO regulators using RC couplings.

Finally, in Chap. 6, some conclusions are drawn with emphasis on the contributions presented in this book.

Seville, Spain
January 2018

José María Hinojo
Clara Luján Martínez
Antonio Torralba

Acknowledgements

We would like to acknowledge the Spanish Ministry of Economy and Competitiveness under project TEC2015-71072-C3-3-R, the Andalousian 'Consejería de Economía, Innovación y Ciencia' under the project P12-TIC-1862, and Universidad de Sevilla, for their support to this research.

We would also like to express great gratitude to all our colleagues in circuit design, especially to Profs. Jaime Ramírez-Angulo and Ramón G. Carvajal for their invaluable advice, suggestions and extremely fruitful technical discussions.

Lastly, we would like to thank all reviewers for their diligence in reviewing this manuscript as well as the staff of Springer for their support and encouragement to publish this work.

Contents

Acronyms

CAFVF	Cascode Flipped Voltage Follower
CCI	First-Generation Current Conveyor
CDMA	Coded Division Multiple Access
C-FSP	Charging-Fast Settling Path
CMOS	Complementary MOS
DC	Direct Current
DF	Damping Factor
D-FSP	Discharging-Fast Settling Path
DFVS	Dynamic Frequency and Voltage Scaling
DMM	Digital Multimeter
DRAM	Dynamic RAM
DVS	Dynamic Voltage Scaling
EA	Error Amplifier
EC-LDO	Externally Compensated Low-Dropout
EEPROM	Electrically Erasable Programmable ROM
ESR	Equivalent Series Resistance
FOM	Figure of Merit
FVF	Flipped Voltage Follower
HCFC	Hybrid Cascode Feedforward Compensation
HDO	High-Dropout
IC-LDO	Internally Compensated Low-Dropout
IoT	Internet of Things
KCL	Kirchhoff's Current Law
LCD	Liquid Crystal Display
LDO	Low-Dropout
LHP	Left Half-Plane
MOS	Metal-Oxide Semiconductor
NMC	Nested Miller Compensation
OTA	Operational Transconductance Amplifier
PCB	Printed Circuit Board

PDK	Process Design Kit
PM	Phase Margin
PSRR	Power Supply Ripple Rejection
PWM	Pulse Width Modulation
RAM	Random Access Memory
RF	Radio Frequency
RFID	Radio-Frequency Identification
RHP	Right Half-Plane
RNMC	Reverse Nested Miller Compensation
ROM	Read-Only Memory
SC	Switched Capacitor
SoC	System-on-Chip
SR	Slew-Rate
SRAM	Static RAM
STUR	Subtreshold Undershoot Reduction
UGF	Unity Gain Frequency
VCO	Voltage Controlled Oscillator
VNA	Vector Network Analyzer
WSN	Wireless Sensor Network
ZNR	Zero-Nulling Resistor

List of Figures

List of Tables

Chapter 1
Introduction

Abstract The continuous downscaling of semiconductor technology, predicted by Gordon Moore in 1965, has had a major impact on the development of integrated electronics. The reduction of transistor size has allowed the integration of more and more devices in the same die, increasing the integration density. In addition, it has led to the reduction of fabrication costs, making the final product cheaper and more accessible. However, this increase in the functionality of an integrated circuit entails greater complexity in the generation and distribution of the different supply voltages required in a chip. As more systems are integrated into the same die, more biasing domains coexist requiring different noise, regulation and/or stability specifications to be simultaneously satisfied. Therefore, power management circuits have been acquiring greater significance as technology downscales, reaching a maximum at present, in the nanoscale era. Voltage regulators are key components for power management, as they provide a regulated, stable and noise-free supply voltage to the active blocks of an SoC. This chapter reviews the basics of voltage regulation with emphasis on linear regulators.

1.1 Introduction

Battery-powered devices are gaining relevance. The development of the so-called Internet of Things (IoT) [1–4] is revolutionizing the conception that human beings have about the world and the way of interacting with it, both in the personal and urban areas, as well as in the industry. This new interconnection paradigm has led to a widespread use of solutions based on low-power wireless networks [5–8], whose autonomous sensor nodes measure different magnitudes and transmit them to other nodes in the network, and to the rest of the world. These nodes have strict requirements in terms of battery lifetime, which, in some cases, exceeds 10 years. In addition, alongside the mobile telephony, some other fields for wireless applications, like health care and well-being, favour the appearance of wearable devices [9].

The downscaling of technologies is one of the main reasons that make possible this trend. According to Moore's law, the number of transistors on a chip approximately doubles every 2 years [10]. Thus, billions of transistors can be found in a modern

© Springer International Publishing AG, part of Springer Nature 2018
J. M. Hinojo et al., *Internally Compensated LDO Regulators for Modern System-on-Chip Design*, https://doi.org/10.1007/978-3-319-75411-6_1

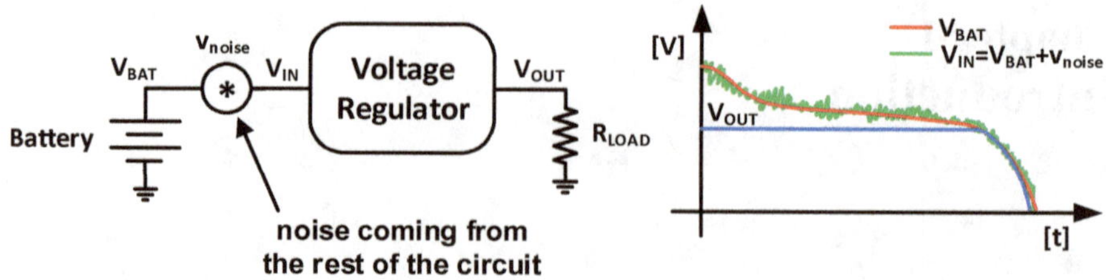

Fig. 1.1 Battery discharge curve for a resistive load, R_{LOAD}, in a noisy environment, and the voltage at the regulator output, V_{OUT}

integrated circuit, which reduces its final cost and increases its complexity, as a large number of different functions can be integrated into the same chip. This has led to the SoC concept. A SoC integrates all the necessary subsystems, whether analogue, radio frequency or digital, to perform the required function on the same substrate. This new paradigm within the semiconductor industry has motivated a change in the way systems are designed. A new subsystem integration scheme is superimposed on the traditional hierarchical method, where a central processor acted as the coordinator.

The high complexity of a SoC has also changed the priority in the design objectives for integrated circuits. During the nineties, the priority was focused on improving operating performances, while aspects such as low power consumption or low supply voltage were of less importance. The growing demand for SoCs and the need to increase the battery lifetime in autonomous devices [11] have made low power consumption and energy efficiency the new priorities.

Modern SoCs require stable, precise and well-regulated supply voltages. This has caused power management systems, and specifically voltage regulators, to become critical cells. Regulators convert the supply voltage, which could come from a noisy voltage source of a different value, even varying in time, to a regulated, precise and stable output voltage. Figure 1.1 depicts the complexity and challenges that these regulators face. An example that highlights the current importance of voltage regulators and power management blocks can be found in [12], where the power supply requirements for a Code-Division Multiple Access (CDMA) modem of a mobile phone are described. Up to 11 independent supply voltages are required to ensure proper operation. Furthermore, the modem integrates a control block that minimizes energy consumption by supervising each subsystem, switching any of them off when its operation is not required.

1.2 Voltage Regulators

Devices responsible for supplying and conditioning power can be classified into three main categories. The first category comprehends those which are called Switched-Capacitor (SC) Regulators. They are based on charge transfer between capacitors to convert the input voltage in a regulated output voltage. The second group is known as switching regulators. They make use of an inductor for intermediate storage of the energy that is, then, transferred to the load. Finally, the last category is named as linear regulators because they are based on an active element in series with the load that dynamically adapts the voltage between its terminals in order to supply a regulated output voltage.

1.2.1 Switched-Capacitor Regulators

SC regulators are also known as charge pumps because they are similar to a pump that forces the charge to flow from the input to the output, using capacitors for intermediate energy storage. They are extensively used in integrated circuits due to their high versatility. These regulators are able to step up or down the input voltage, or even to change its polarity. They are used in those applications whose overall requirements are not highly demanding and thus, neither high output power nor high efficiency is required. Non-volatile memories such as flash memories, Electrically Erasable Programmable Read-Only Memories (EEPROMs) or Dynamic Random Access Memories (DRAM) are examples where this kind of regulators is applied. Another possible use is in Liquid Crystal Displays (LCD) to generate the backlight, in diode emitters to generate their bias voltage, in Radio-Frequency Identification (RFID) tags, and more recently, in energy harvesting circuits, where they are able to boost the voltage generated by transducers (e.g. a solar cell, a thermoelectric generator or a piezoelectric sensor) to an output voltage that supplies the rest of the system.

As it can be observed in Fig. 1.2, SC regulators have a set of capacitors that exchange their charge when a set of switches are opened and closed. For the sake of simplicity, this figure depicts one SC stage with only two capacitors, C_1 and C_{LOAD}. In addition, the clock sequence that alternatively opens and closes the switches is also represented in this figure. During clock phase φ_1, switches labelled S_1 are closed while switches labelled S_2 are open. In this time interval, the voltage across capacitor C_1 is, in the steady state, equal to the input voltage V_{IN}. Then, in the clock phase φ_2, switches S_1 are open and S_2 are closed. Now, C_1, initially charged to $Q_{C1} = C_1V_{IN}$, is in series with the input voltage source. Assuming no C_{LOAD}, at the end of phase φ_2, the output voltage is given by $V_{OUT} = 2V_{IN}$, stepping the input voltage up. If an uncharged C_{LOAD} is placed at the output node, assuming ideal components, the output voltage will be given by Eq. 1.1.

$$V_{OUT} = \frac{C_1}{C_1 + C_{LOAD}} 2V_{IN} \tag{1.1}$$

In the previous example, in the absence of C_{LOAD}, the input voltage is multiplied by a factor of 2. Other factors can be obtained with other, usually more complex, topologies. Some of these more complex schemes are proposed in [13–20]. As it

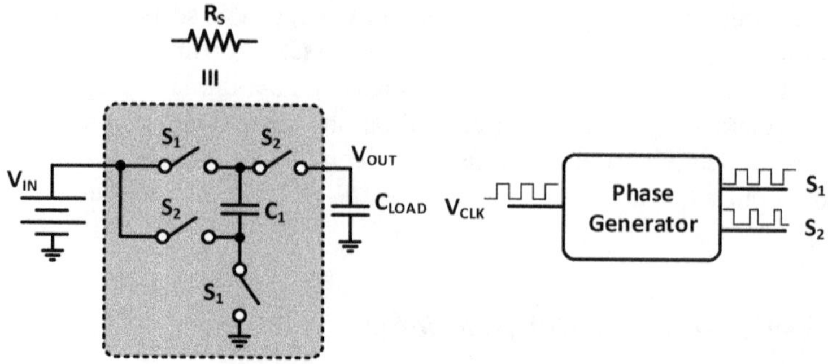

Fig. 1.2 Block diagram of a switched-capacitor regulator

can be concluded from these references, SC regulators are highly versatile as, by modifying the switching matrix, different output voltages can be generated.

The main disadvantage of this kind of regulators is their limited capability for maintaining the output voltage for a wide range of load currents. When the load current changes, charge pumps are required to adapt their equivalent resistance, R_S, to the new load condition. In order to modify R_S, either, the switching frequency, or the capacitance of floating capacitors should be charged. In both cases, the efficiency of the charge pump changes [18, 21]. In practice, designers tend to keep constant the equivalent resistance and change the conversion ratio, which is defined as the relation between V_{IN} and V_{OUT}, to maximize efficiency. Under this condition, the voltage V_{OUT} is bound by upper and lower limits. Beyond these limits, the regulated voltage V_{OUT} can be improved at the expense of increasing the number of floating capacitors or the complexity of the switching scheme. Therefore, SC regulators are implemented in circuits without stringent requirements concerning current load variations.

In a practical implementation of an SC regulator, the power efficiency is limited by several constraints. To charge capacitors to their final value, the time constant RC has to be much smaller than the switching period. Note that the capacitive term is defined by the floating capacitor, whereas the resistance is given by the on resistance of switches, R_{ON}, and the equivalent series resistance, ESR, of capacitors. Provided the capacitor ESR is small enough, a large transistor aspect ratio must be chosen for switches to have a small R_{ON}. Large switches mean large parasitic capacitances, which add up to the parasitic capacitance of the floating capacitor. As a consequence, the circuitry that drives the switches increases its power consumption, especially if a high switching frequency is chosen. Stratakos and others [22] proposed a methodology to find an optimal size for switches used in DC-DC converters, selecting conduction losses to be equal to gate-driving losses. A second issue to be considered is the conversion ratio, which depends on the circuit topology. To achieve a given conversion ratio, several floating capacitors and their corresponding switching matrix might be necessary, which would increase the silicon area.

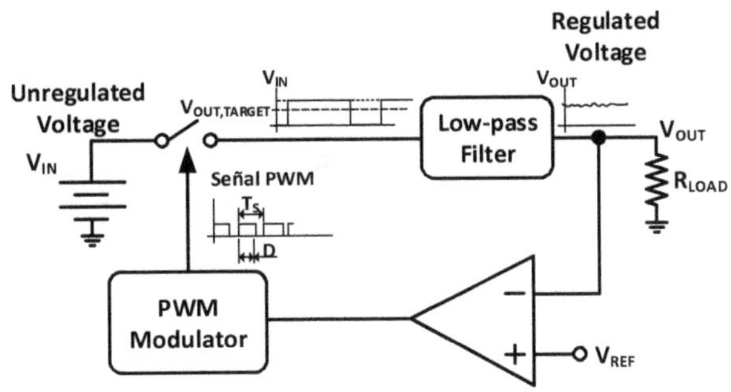

Fig. 1.3 Block diagram of a switching regulator

To maximize the power efficiency, parasitic capacitances can be reduced by selecting expensive CMOS technologies, such as Silicon-on-Insulator (SOI). However, a standard CMOS process is a common requirement in the design of LDO regulators.

1.2.2 Switching Regulators

Switching regulators are composed of a transistor operating as a switch, a low-pass filter and a control block. Figure 1.3 shows the block diagram of a switching regulator. In these regulators, the switch controls the amount of unregulated voltage delivered to the output voltage, V_{OUT}, through pulses. Next, the low-pass filter cuts off the high-frequency components of these pulses in order to provide a steady value for V_{OUT}. In addition, there is a control block that implements the regulation loop. This block senses the averaged output voltage and compares it to a reference, V_{REF}, in order to select a proper value for the width of the power pulses. In this way, the mean value of V_{OUT} is controlled by Pulse Width Modulation (PWM).

Unlike SC regulators, the energy efficiency of switching regulators can be high. In actual implementations, these regulators achieve efficiency between 75 and 90% [23–28]. Many different topologies for the switching and filtering blocks have been used. In spite of the high efficiency, their main disadvantage comes from the need of an inductor, which simultaneously serves as an intermediate energy storage element, and as part of the output filter. To attain a high efficiency in a fully integrated regulator, an expensive technology with high quality-factor inductors is required. For that reason, off-chips inductors are used, at the cost of increasing the number of components in the PCB. The effect of the parasitic elements introduced by the package should also be considered in this case.

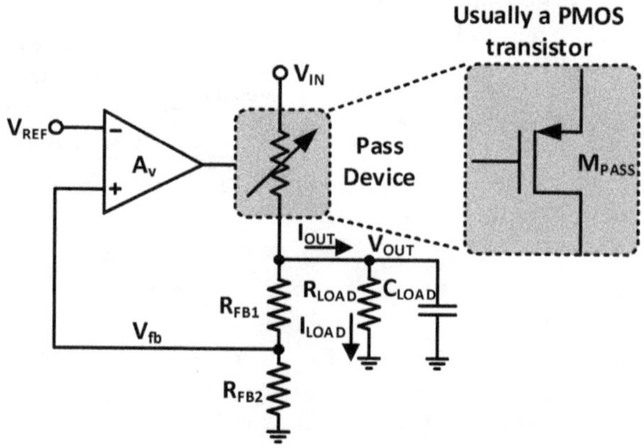

Fig. 1.4 Block diagram of a classical linear regulator

1.2.3 Linear Regulators

Linear regulators are the counterparts to switching regulators. The classical topology for these circuits is represented in Fig. 1.4, where three main blocks can be observed. The first component is the pass element placed in series with the load, usually implemented by a transistor (M_{PASS}). It supplies the current required by the load. Next block is the feedback network, which is usually implemented by a resistor divider that provides a scaled version of the output voltage, V_{OUT}, to be compared to the reference voltage, V_{REF}. Finally, the last element is the so-called Error Amplifier (EA) that compares the scaled version of V_{OUT} to V_{REF}, and modulates the controlling voltage of the pass element, so that V_{OUT} is kept constant, regardless the output load and input voltage.

In linear regulators, there is an important parameter that not only determines the minimum unregulated input voltage required for proper functioning but also limits the minimum power dissipated by the regulator: the dropout voltage, $V_{dropout}$. This parameter is defined as the difference between the input voltage from the unregulated power source, V_{IN}, and the regulated output voltage V_{OUT}. According to the dropout voltage, linear regulators can be classified into High-Dropout (HDO) ones, when the dropout voltage is higher than 0.6 V, and Low-Dropout (LDO) ones otherwise. This last group typically presents a dropout voltage between 0.15 and 0.35 V [29]. Therefore, LDO regulators dissipate less power than HDO ones. This makes LDO regulators to become popular cells in the increasingly demanded market of portable and handled devices.

Usually, LDO regulators suffer from stability problems and require compensation. According to the selected compensation, they are classified in externally (EC) or internally (IC) compensated regulators. In the first case, a large capacitor is connected at the output of the linear regulator in order to force the location of the dominant pole at this node. A typical value is between 1 to 10 μF. Conversely, an IC-LDO regulator makes use of an internal node to place the dominant pole. Usually, this node is associated to the gate of M_{PASS}. Due to its large size, the total parasitic capacitance at the gate of M_{PASS}, C_{GATE}, is usually in the order of tens of picofarads. The first non-dominant pole of IC-LDO regulators is fixed by the output impedance of the linear regulator, Z_{OUT}, which is due to R_{OUT} and C_{OUT}, where R_{OUT} is the equivalent output resistance (including R_{LOAD}) and C_{OUT} is the total output capacitance, $C_{OUT} = C_{LOAD} + C_{par}$, and C_{par} is the parasitic capacitance at the regulator output. In IC-LDO regulators, C_{OUT} is dominated by the load capacitor, so that $C_{OUT} \approx C_{LOAD}$. Z_{OUT} needs to be smaller than a given maximum value so as not to degrade the regulator stability.

As shown in Fig. 1.4, an IC-LDO regulator is a simple cell, as it only requires a few elements to provide a well-regulated output voltage. Therefore, it is possible to achieve implementations with small silicon area and low power consumption. Nevertheless, there are also some drawbacks that should be considered. First, the unregulated input voltage V_{IN} must be greater than the regulated output V_{OUT}. Second, the power efficiency η presents a theoretical upper bound imposed by the ratio $V_{OUT}/V_{IN,min}$.

The power efficiency is defined as

$$\eta = \frac{P_{OUT}}{P_{IN}} = \frac{V_{OUT} \cdot I_{OUT}}{V_{IN} \cdot I_{IN}} = \frac{V_{OUT} \cdot I_{OUT}}{V_{IN} \cdot (I_{OUT} + I_Q)} \tag{1.2}$$

where I_Q is the LDO quiescent current consumption, and I_{OUT} is the DC component of the current provided by M_{PASS} to the load. Taking into account that higher the input voltage, higher the dropout voltage, it is straightforward to conclude that the power efficiency is maximized when the input voltage V_{IN} is close to its minimum value (Eq. 1.3) and the quiescent current consumption is reduced as much as possible ($I_Q \approx 0$).

$$V_{IN,min} = V_{OUT,nom} + V_{dropout} \tag{1.3}$$

An example of how the efficiency of a linear regulator is modified when the current load changes is shown in Fig. 1.5, where the efficiency is depicted for different values of the input voltage and four different values of the quiescent current consumption. These curves have been obtained evaluating Eq. 1.2 for $V_{OUT} = 1$ V and $V_{dropout} = 0.2$ V. As it can be observed, the power efficiency is upper bounded by Eq. 1.4.

$$\eta = \frac{V_{OUT} \cdot I_{OUT}}{V_{IN} \cdot (I_{OUT} + I_Q)} \le \frac{V_{OUT}}{V_{IN,min}} = \frac{V_{OUT}}{V_{OUT} + V_{dropout}} \tag{1.4}$$

The maximum of the efficiency, η_{max}, can be approximated as

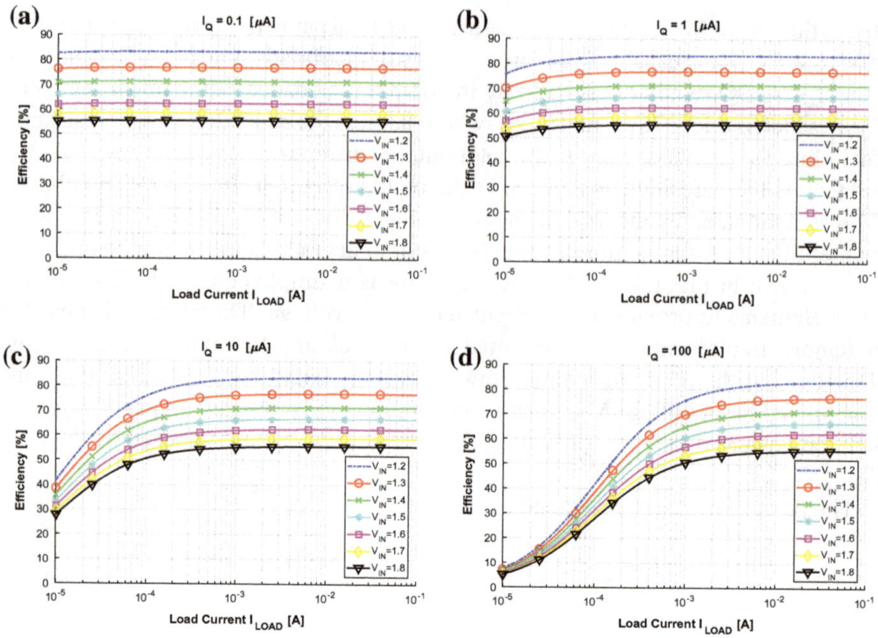

Fig. 1.5 Power efficiency versus variations in the load current for different values of the input voltage and quiescent current consumption

$$\eta_{max} = \frac{V_{OUT}}{V_{OUT} + V_{dropout}} = \frac{1}{1 + V_{dropout}/V_{OUT}} \approx 1 - \frac{V_{dropout}}{V_{OUT}} + \left(\frac{V_{dropout}}{V_{OUT}}\right)^2 \tag{1.5}$$

assuming $V_{dropout} \ll V_{OUT}$. For $V_{OUT} = 1$ V and $V_{dropout} = 0.2$ V, the above expression leads to a maximum efficiency of about 84 %, which is the approximate limit for the efficiency in the curves of Fig. 1.5. Concerning Fig. 1.5, note that, in DC, $I_{LOAD} = I_{OUT}$.

1.2.4 Comparison

Table 1.1 portrays the main performances of the different types of regulators that have been previously presented, in order to ease their comparison. In this table, it can be observed how the major advantage of linear regulators is a low noise, well-regulated output voltage. Moreover, it is possible to save PCB area and complexity, as they require fewer external components than their switching counterparts. Nonetheless, linear regulators present two major drawbacks. The first one is an efficiency lower than that of switching regulators, although when the unregulated input voltage is close to the output voltage (i.e. the dropout voltage is small) their efficiency can still be high. The second one stems from the fact that the output voltage can never be

Table 1.1 Comparison of main performances and usual applications for linear, switched-capacitor and switching voltage regulators

Regulator topology	Linear	Switching capacitor	Switching
Boost	No	Yes	Yes
Buck	Yes	Yes	Yes
Negative V_{OUT}	No	Yes	Yes
Efficiency	Poor	Poor	High
Noise	Low	High	High
Regulation	High	Poor	Medium
Area	Small	Medium/Large	Medium/Large
Cost	Low	Medium	High
Applications	Analogue cells, RF, Large regulation	DRAM, Flash, EEPROM	Microprocessors, Digital cells, SRAMs

greater than the unregulated input voltage. In any case, linear regulators are massively implemented in high-performance applications where a fast transient response to variations in the input voltage and the current load is required, as well as a high rejection to power supply perturbations.

1.3 Constraints and Challenges for IC-LDO Regulators

LDO regulators, and specifically those which are internally compensated, are currently highly demanded cells in the industry and a hot topic in the research community. Present wireless applications demand several orders of magnitude reduction in the power consumption of battery-operated devices. Thus, LDO regulator designers must face the challenge of a well-regulated output voltage with low ripple and fast transient response under low power constraint.

First of all, the transient response of an IC-LDO regulator is seriously affected in a low-power environment, as a large current is required to charge or discharge the large parasitic capacitance of the pass transistor. With current sources of a small value, the output voltage may present a high ripple, which is unacceptable in many applications such as biomedical front ends or RF transceivers.

Second, in a low-power environment, the frequency response of systems is usually limited by the presence of low and medium frequency poles, which degrade the stability of the system. In an IC-LDO regulator, this issue is even more critical, as it must deal with large variations in the unregulated input voltage, load current and output capacitance. It is difficult to stabilize LDO regulators under such varying external conditions.

Finally, the regulator must be also able to suppress the noise caused by other switching regulators that could be connected at the output of the LDO regulator. Therefore, an adequate Power Supply Ripple Rejection (PSRR) is required. This issue becomes especially significant in IC-LDO regulators as a consequence of the low value of the output impedance. As it will be shown in a later chapter, the PSRR response is dominated by a pole, caused for the impedance at the gate of M_{PASS}, and a zero generated by C_{LOAD}. Therefore, the PSRR response rapidly goes to zero decibels, coupling the high frequency variations of the unregulated input voltage to V_{OUT}.

1.4 Contributions of This Book

As stated before, an IC-LDO regulator must be able to regulate the input voltage, minimizing the variations of V_{OUT} to changes in load current, input voltage and load capacitance. This book aims to overview the design of IC-LDO regulators under low power and low voltage constraints. In the remaining chapters, other considerations will be made about stability, transient response and new topologies of IC-LDO regulators.

In addition, some innovative contributions will be presented.

- **Contribution 1**. A novel Miller-based compensation scheme is proposed to improve the stability of IC-LDO regulators. It employs a replica circuit that senses the operating region of the pass transistor and generates a control voltage that changes the value of the zero-nulling resistor. This scheme is shown to extend the stability of Miller-compensated IC-LDO regulators to a wide range of load currents.
- **Contribution 2**. In order to minimize the quiescent consumption and enhance the transient response, a class AB buffer is inserted to drive the parasitic capacitance of the pass transistor.
- **Contribution 3**. A Flipped-Voltage-Follower- (FVF-) based IC-LDO regulator is proposed. The FVF cell represents a simple way to implement the error amplifier and provides low output impedance with very low quiescent power consumption.

The proposed regulators in these contributions have been designed and fabricated in standard 180-nm (contribution 2) and 65-nm (contributions 1 and 3) CMOS technologies. The experimental results reported in different chapters of this book show that they are in, or close to, the state of the art.

Chapter 2
Internally Compensated LDO Regulators

Abstract This chapter contains an introduction to Internally Compensated Low-Dropout (IC-LDO) regulators. The design of these circuits and the most used Figures of Merit (FOMs) to evaluate their performances are studied. Special attention is paid to three aspects of their design: (a) stability. In IC-LDO regulators, the dominant pole is located at an inner node, while the non-dominant pole, located at the output, is responsible for the degradation of the stability. Furthermore, it depends on the load condition, which complicates the design of a compensation network. (b) Transient response: The regulator load usually requires fast transient response to load current and input voltage variations, and (c) power supply ripple rejection. Perturbations in the input voltage cause undesired disturbances in the output voltage. This chapter discusses the techniques proposed in the literature to face these design challenges, with emphasis on low power solutions. In addition, and based on a set of selected figures of merit, a comparison of recently published LDO regulators is made at the end of the chapter.

2.1 Some Key Concepts

Figure 2.1 shows the classical topology of an LDO regulator where the pass element is implemented by a PMOS (Fig. 2.1a) or NMOS (Fig. 2.1b) transistor, whose dropout voltage can be as low as a one $V_{SD,sat}$. $V_{SD,sat}$ is the minimum source-to-drain voltage to keep the transistor in saturation region and can be as low as 0.1 V. R_{LOAD} and C_{LOAD} model the load resistance and capacitance, respectively, while R_{FB1} and R_{FB2} are the resistors of the feedback network. In the case of battery-powered devices, where the power is a scarce resource, designers have to maximize the battery lifetime as much as possible. In order to achieve this goal, a PMOS transistor is more advantageous than an NMOS one, as the minimum input voltage to properly work is $V_{IN,min} = V_{OUT} + V_{SD,sat}$. However, for an NMOS regulator, it is $V_{IN,min} = V_{OUT} + V_{DS,sat} + V_{GS}$. As a consequence, a regulator based on a PMOS pass transistor has a wider range of operation. On the other hand, it occupies more area for the same current, due to its low carrier mobility.

© Springer International Publishing AG, part of Springer Nature 2018
J. M. Hinojo et al., *Internally Compensated LDO Regulators for Modern System-on-Chip Design*, https://doi.org/10.1007/978-3-319-75411-6_2

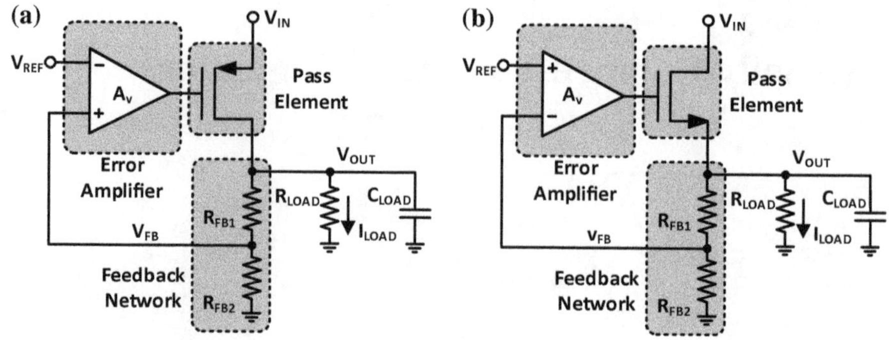

Fig. 2.1 Classical topology of an LDO regulator: **a** PMOS-type and **b** NMOS-type

Fig. 2.2 Small-signal model of a PMOS-type LDO regulator

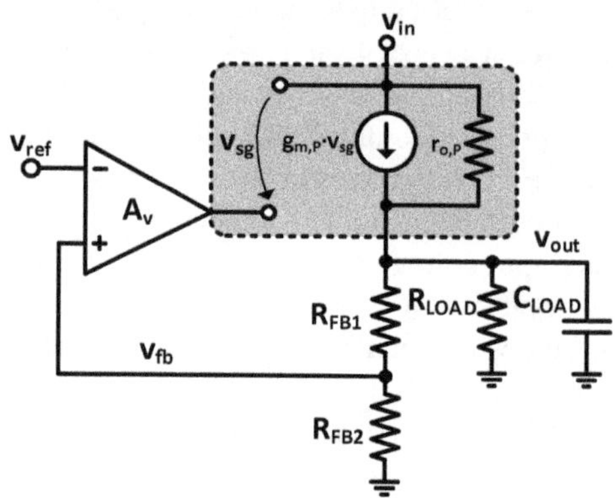

Although load and line variations affect the large signal behaviour of the regulator, a first-order estimation can be made using its small-signal model and assuming an ideal error amplifier (Fig. 2.2).

Applying Kirchhoff's current law in the output node for frequencies near DC:

$$g_{m,P}\left[v_{in} - A_v\left(\frac{R_{FB2}}{R_{FB1}+R_{FB2}}v_{out} - v_{ref}\right)\right] + \frac{v_{in}-v_{out}}{r_{o,P}} = \\ = v_{out}\left(\frac{1}{R_{FB1}+R_{FB2}} + \frac{1}{R_{LOAD}}\right) \tag{2.1}$$

$$v_{out} = \frac{g_{m,P} + \frac{1}{r_{o,P}}}{g_{m,P}A_v\frac{R_{FB2}}{R_{FB1}+R_{FB2}} + \frac{1}{(R_{FB1}+R_{FB2})||R_{LOAD}||r_{o,P}}}v_{in} + \\ + \frac{g_{m,P}A_v}{g_{m,P}A_v\frac{R_{FB2}}{R_{FB1}+R_{FB2}} + \frac{1}{(R_{FB1}+R_{FB2})||R_{LOAD}||r_{o,P}}}v_{ref} \tag{2.2}$$

where $g_{m,P}$ and $r_{o,P}$ are the transconductance and output resistance of M_{PASS}, respectively, and A_v is the gain of the error amplifier.

Assuming $g_{m,P}r_{o,P} \gg 1$ and $g_{m,P}A_v\left[(R_{FB1}+R_{FB2})\,\|R_{LOAD}\|r_{o,P}\right]\cdot \cdot\frac{R_{FB2}}{R_{FB1}+R_{FB2}} \gg 1$, Eq. 2.2 can be simplified to

$$v_{out} \approx \frac{g_{m,P}r_{o,P}}{g_{m,P}r_{o,P}A_v\frac{R_{FB2}}{R_{FB1}+R_{FB2}}}v_{in} + \frac{1}{\frac{R_{FB2}}{R_{FB1}+R_{FB2}}}v_{ref} \tag{2.3}$$

Let β be the gain of the feedback network, $\beta = \frac{R_{FB2}}{R_{FB1}+R_{FB2}}$; then, Eq. 2.3 can be reduced to Eq. 2.4. Note that the error amplifier attenuates the variations in the input voltage v_{in}, but it has no effect on the reference voltage v_{ref}.

$$v_{out} \cong \frac{1}{A_v\beta}v_{in} + \frac{1}{\beta}v_{ref} \tag{2.4}$$

2.1.1 Line Regulation

Line regulation measures the variation of the output voltage caused by a change in the steady-state input voltage, and it is defined as

$$Line\ Reg. = \frac{\Delta V_{OUT}}{\Delta V_{IN}} \tag{2.5}$$

In the ideal case, the output voltage does not depend on the input voltage, so that the value of the line regulation should be zero. However, the finite gain of the error amplifier causes a dependence of V_{OUT} with V_{IN}.

Using the small-signal model of Fig. 2.2 with $v_{ref} = 0\,V$, and assuming a small-signal change at the input v_{in}, from Eq. 2.5, the line regulation can be approximated by

$$Line\ Reg. = \frac{\Delta V_{OUT}}{\Delta V_{IN}} \approx \frac{1}{A_v\beta} \tag{2.6}$$

According to Eq. 2.6, the larger the error amplifier gain, the better the line regulation.

2.1.2 Load Regulation

Load regulation measures the variation of the output voltage caused by a change in the steady-state load current. It is defined as

$$Load\ Reg. = \frac{\Delta V_{OUT}}{\Delta I_{LOAD}} \quad (2.7)$$

Using, once again, the small-signal model of Fig. 2.2 with $v_{in} = v_{ref} = 0$ V, for a change in the load current i_{load}, the load regulation can be approximated by

$$Load\ Reg. = \frac{\Delta V_{OUT}}{\Delta I_{LOAD}} \approx -\frac{1}{g_{m,P} A_v \beta} \quad (2.8)$$

According to Eq. 2.8, the load regulation not only improves when the error amplifier gain increases but also with the transconductance of M_{PASS}. This last statement implies that M_{PASS} should operate in saturation region. However, for those regulators that must provide high I_{LOAD}, M_{PASS} is usually forced to work at maximum load current in the triode region, to save area.

2.2 Issues Related to IC-LDO Regulators

2.2.1 *Stability Analysis*

Regarding frequency response, LDO regulators must operate in a wide range of conditions, where its stability must be guaranteed. An uncompensated LDO regulator has two major poles. One of them is determined by the large parasitic capacitance at the gate of M_{PASS}, C_{GATE}. The second pole is located at the output node, and it depends on the load conditions. The rest of internal poles introduced by the error amplifier can be assumed to be, approximately, high-frequency poles and, therefore, in a first-order approach, their effect on the system can be neglected. Assuming that the small-signal model of Fig. 2.3 represents an uncompensated LDO regulator, these two poles are given by Eqs. 2.9 and 2.10, respectively. In addition, as a consequence of the parasitic capacitance $C_{gd,P}$, a Right Half-Plane (RHP) zero, whose expression is given by Eq. 2.11, appears beyond the Unity Gain Frequency (UGF).

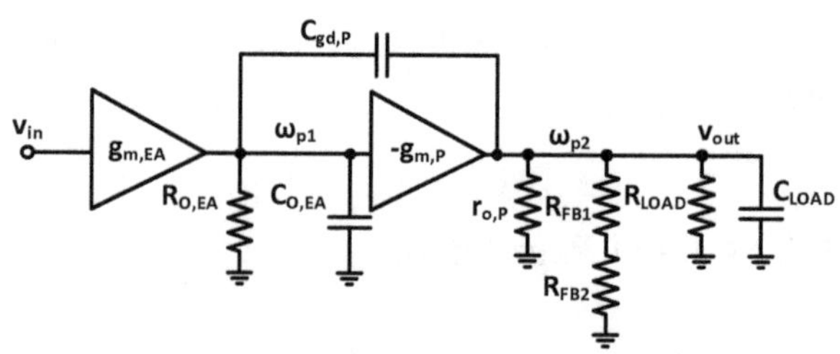

Fig. 2.3 Small-signal model of an uncompensated LDO regulator

Table 2.1 Table with the values used to obtain the curves in this Chapter

Magnitude	Value	Units
A_v	10^6	[V/V]
R_{FB1}	10	[kΩ]
R_{FB2}	10	[kΩ]
$R_{O,EA}$	1	[kΩ]
C_{GATE}	25	[pF]
$(W/L)\vert M_{PASS}$	$250 \times (50/0.18)$	[μm]/[μm]

In the design of an IC-LDO, ω_{p1} is intended to be the dominant pole.

$$\omega_{p1} \approx -\frac{1}{R_{O,EA}\left(C_{O,EA} + g_{m,P}R_{OUT}C_{gd,P}\right)} \tag{2.9}$$

$$\omega_{p2} \approx -\frac{1}{R_{OUT}C_{LOAD}\frac{C_{O,EA}+C_{gd,P}}{C_{O,EA}+g_{m,P}R_{OUT}C_{gd,P}}} \tag{2.10}$$

$$\omega_{z1} \approx \frac{g_{m,P}}{C_{gd,P}} \tag{2.11}$$

In this model, Fig. 2.3, $g_{m,EA}$, $R_{O,EA}$ and $C_{O,EA}$ are the equivalent transconductance, the output resistance and the output capacitance of the error amplifier, respectively. At the same time, R_{OUT} is the equivalent total output resistance of the LDO regulator, which includes the effect of the load current modelled by R_{LOAD}, and is given, approximately by $R_{OUT} = r_{o,P}||R_{LOAD}||(R_{FB1}+R_{FB2})$. C_{GATE} is the total equivalent parasitic capacitance at the gate of M_{PASS}, $C_{GATE} = C_{gs,P}+C_{O,EA}$, C_{LOAD} models the load capacitance and $g_{m,P}$ represents the transconductance of M_{PASS}. From the previous expressions, a qualitative analysis of the stability can be made as a function of three parameters: I_{LOAD}, V_{IN} and C_{LOAD}. In what follows, when necessary, we will use the typical parameters of a standard 0.18 μm CMOS technology, i.e. $V_{TN} = 0.46$ V and $V_{TP} = -0.43$ V. In addition, Table 2.1 summarizes the design parameters used to obtain the curves in this chapter.

Load Current

In this subsection, the stability of an IC-LDO regulator under extreme variations of the load current is discussed. From the pole-zero model presented above, it is known that the dominant pole ω_{p1}, Eq. 2.9, depends on $g_{m,P}R_{OUT}$, which, in a first-order approach, depends on I_{LOAD}. However, $g_{m,P}R_{OUT}$ remains approximately constant for I_{LOAD} from 0 A up to tens of μA (Fig. 2.4), and it only becomes highly dependent on I_{LOAD} when $R_{OUT} \approx R_{LOAD}$ and M_{PASS} operates in triode region (i.e. for I_{LOAD} ranging from several mA to hundreds of mA).

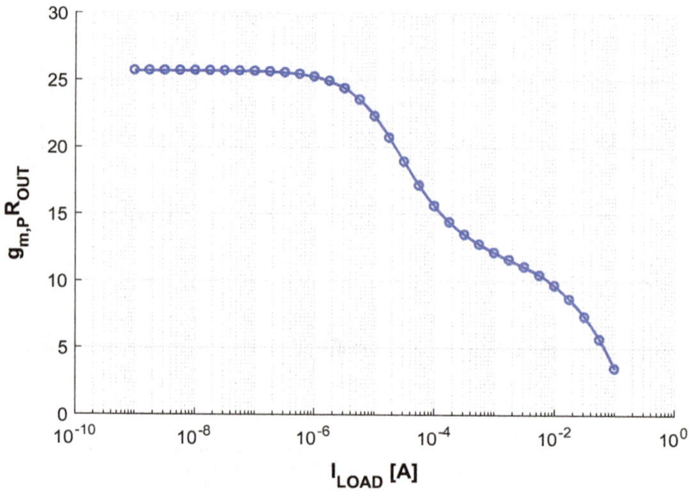

Fig. 2.4 $g_{m,P}R_{OUT}$ value for different I_{LOAD} values

Even in this case, the dominant pole condition is still valid and, thus, the regulator remains stable. Therefore, it can be assumed that, regarding stability, the effect of the load current on ω_{p1} is negligible for the entire range of I_{LOAD}.

Regarding ω_{p2}, its location significantly changes with I_{LOAD}, since it depends not only on the product $g_{m,P}R_{OUT}$ but also, directly, on R_{OUT} (Eq. 2.10). In this example, the value of R_{OUT} is $\sim 1\ \Omega$, for I_{LOAD} in the order of several mA, and increases when I_{LOAD} decreases, approaching a value of 100 kΩ when the load current reaches zero. Consequently, as I_{LOAD} reduces its value, ω_{p2} moves towards the UGF, degrading the Phase Margin (PM) of the entire system. As a consequence, an uncompensated LDO tends to be unstable under small and zero I_{LOAD} conditions. This effect is graphically described in Fig. 2.5.

Finally, the RHP zero also modifies its location. According to Eq. 2.11, ω_{z1} is directly proportional to $g_{m,P}$ and, therefore, it will move towards lower frequencies (close to the UGF) when I_{LOAD} decreases (see Fig. 2.5).

Input Voltage

Similar to the load current, the input voltage can also suffer a large variation. However, its impact on the frequency response is negligible, as its variation only produces minor changes in the small-signal parameters of M_{PASS}.

C_{LOAD} Variation

According to expressions Eqs. 2.9–2.11, only ω_{p2} depends on the load capacitance. When C_{LOAD} increases, the phase margin is reduced, as ω_{p2} gets closer to the UGF. Note that, in any case, the maximum value of C_{LOAD} is limited by the condition $\omega_{p1} \ll \omega_{p2}$.

Stability Enhancement

Usually, IC-LDO regulators use a multistage amplifier so that their pole-zero map is more complex than that of Fig. 2.5.

Many solutions have been proposed in the literature to improve the stability of IC-LDO regulators with small I_{LOAD} values. One of the earliest proposals [30] uses a compensation block to control the damping factor [31, 32]. This improves the stability of the system and increases the bandwidth. As shown in Fig. 2.6, its implementation requires a compensation capacitor, C_C, which is responsible for placing the dominant pole at the inner node, due to the Miller effect, and a damping factor control block. This block, consisting of a negative gain amplifier and a capacitor, C_{DF}, is intended to increase the quality factor of the pair of conjugate poles that appear at high frequency. In addition, it increases the stability of the system, especially for small loads. According to [32], the most suitable values for these components are

$$C_{DF} = C_C = g_{m,1}\sqrt{\frac{8C_{GATE}C_{LOAD}}{g_{m,2}g_{m,P}}} \tag{2.12}$$

$$g_{m,DF} = 4g_{m,1} \tag{2.13}$$

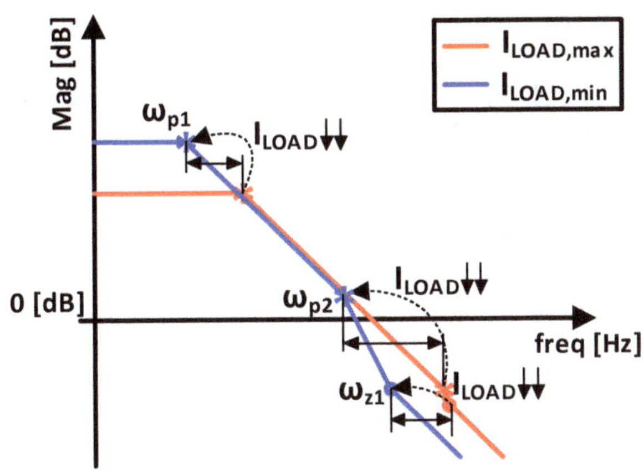

Fig. 2.5 Pole and zero locations in an uncompensated LDO regulator for different I_{LOAD} values

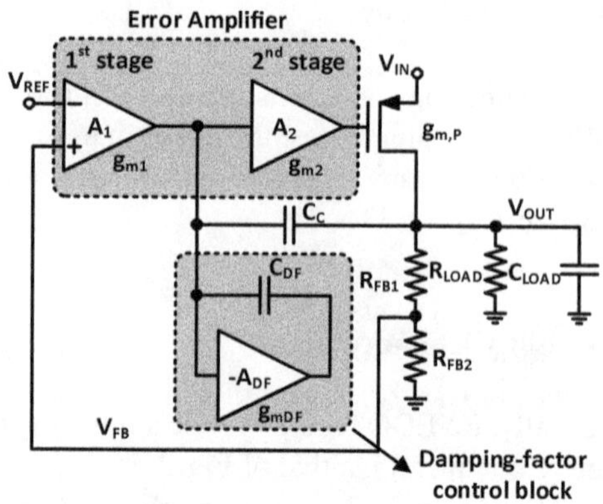

Fig. 2.6 Structure proposed in [30] to improve the stability by controlling the damping factor of the non-dominant poles

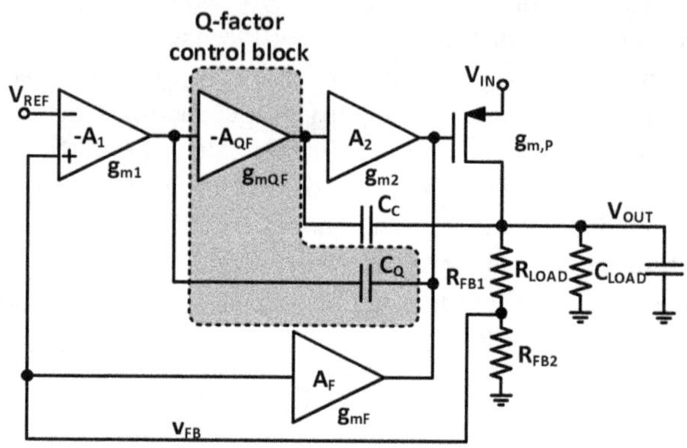

Fig. 2.7 Structure proposed in [33] to improve the stability by controlling the quality factor of the non-dominant poles

where $g_{m,1}$ and $g_{m,2}$ are the transconductance of the first and second stages of the error amplifier, respectively.

When compared to a classical Nested Miller Compensation (NMC), the main advantage of this method is the removal of the inner Miller capacitance, which is located at the first stage of the error amplifier and is the main responsible for the decrease of the regulator bandwidth. Consequently, another advantage is the reduction in area. Its main drawback is the increase in quiescent consumption, due to the introduction of the damping factor control block amplifier, A_{DF}.

A variant of this work can be found in [33], where a block is introduced that controls the quality factor of the pair of non-dominant complex poles. Figure 2.7 shows the corresponding small-signal model. To save power, the active load of the differential pair of the error amplifier is reused as a current buffer. An additional branch is included to introduce a zero in the negative real half-plane with the twofold objective of improving the stability and increasing the maximum current at the gate of M_{PASS}. Unfortunately, every stage of the control circuit is loaded by compensating capacitors, which causes a decrease in the Slew-Rate (SR) of the regulator.

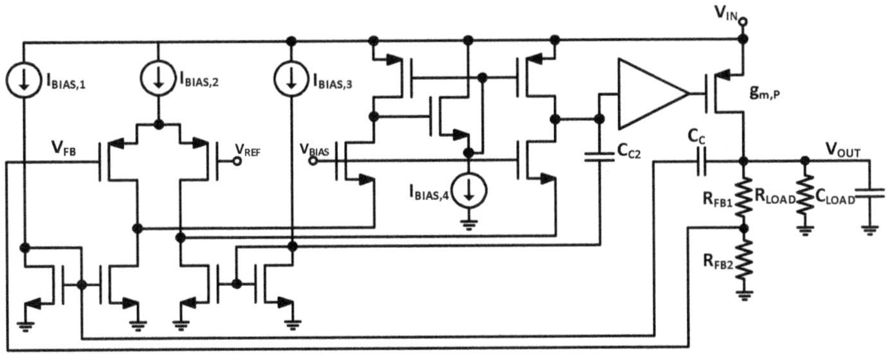

Fig. 2.8 Structure proposed in [36] that uses a capacitive multiplier to reduce the size of the compensating capacitor

Capacitive multipliers based on current buffers have also been used to reduce the size of the compensation capacitors in IC-LDO regulators [34]. It is noteworthy that the area consumed by the active elements in [34] is negligible, as authors reuse some of the blocks already existing in the amplifier to implement a current buffer. However, sometimes it is not possible to reuse existing blocks, and the current buffer has to be implemented with new circuitry, which increases the quiescent consumption [35].

Figure 2.8 shows an example of a capacitive multiplier based on current buffers [36]. This regulator behaves like a two-stage amplifier. The capacitor C_C is responsible for splitting the poles, while C_{C2} reinforces the dominant role of the inner pole. Note that, in addition to the multiplying effect of C_C, the author introduces an intermediate stage with the objective of increasing the slew-rate at the gate of M_{PASS}, and pushing the non-dominant poles at a high frequency.

Capacitive multipliers were also used in [37–44]. As an example, in [37], a differentiator, formed by capacitor C_{CB} and a current buffer (M_1 to M_4), is introduced (Fig. 2.9). This buffer serves a double purpose. First of all, it introduces a fast path between the output of the regulator and the gate of M_{PASS}.

In this way, the variations at the output voltage V_{OUT}, transformed into current by the capacitor C_{CB}, are injected into the gate of M_{PASS} with the objective of increasing the slew-rate of this node, contributing to a better transient response. Second, the buffer helps to separate the poles, since the capacitor C_{CB} appears at M_{PASS} gate multiplied by the gain of the current buffer.

It is worth noting that the use of a current buffer is compatible with other compensation techniques. As an example, in [44], a current buffer is used as part of a classical Reverse Nested Miller Compensation (RNMC).

As a summary of this section, Table 2.2 shows a comparison of the performances of a set of IC-LDO regulators that use some of the compensation techniques discussed above.

Table 2.2 Comparison of selected techniques proposed to improve the stability of LDO regulators

		[30]	[33]	[37]	[40][a]	[41][a]	[42]	[44]
Process	[μm]	0.6	0.35	0.35	0.35	0.18	0.35	0.5
V_{IN}	[V]	1.5–4.5	1.2–3.3	3.0–4.0	1.2	1.1–1.5	1.2–1.5	1.4–4.2
V_{OUT}	[V]	1.3	1.0	2.8	1.0	1.0	1.0	1.21
$V_{DROPOUT}$	[mV]	200	200	200	200	100	200	200
$I_{LOAD,max}$	[mA]	100	100	50	100	50	50	100
I_Q[b]	[μA]	38	–[d]	65	39[c]	54	45	45
C_{OUT}	[pF]	1e2	1e2	1e2	1e4	1e2	1e3	1e5
$\eta\|_{I_{LOAD,max}}$	[%]	99.962	–[d]	99.935	99.961	99.946	99.955	99.955
Area	[mm^2]	0.307	–[d]	0.350	–[d]	–[d]	0.4	0.263
Response time[b]	[μs]	2	50	15	25[c]	2	4	5[c]
ΔV_{OUT} varying V_{IN}								
• Maximum	[mV]	160	–[d]	90	–[d]	–[d]	–[d]	23
• Minimum	[mV]	−1.5	–[d]	−10	–[d]	–[d]	–[d]	−12
$\Delta V_{IN}/t_r$[e]	[$V/\mu s$]	3/6	–[d]	1/1	–[d]	–[d]	–[d]	1/0.1
ΔV_{OUT} varying I_{LOAD}								
• Maximum	[mV]	100	40[c]	80	210	100	70	47
• Minimum	[mV]	−90	−20[c]	−80	−210	−80	−70	−48
$\Delta I_{LOAD}/t_r$[e]	[$mA/\mu s$]	90/0.5	99.9/2[c]	50/1	99/1	50/1	49/1	99.999/0.1
Line Regulation	[mV/V]	–[d]	0.344	–[d]	–[d]	–[d]	0.098	–[d]
Load Regulation	[$\mu V/mA$]	–[d]	338	–[d]	–[d]	–[d]	250	408[c]

[a]Simulation results, [b]Worst case, [c]Estimation based on published results, [d]Not available, [e]t_r: Rise time

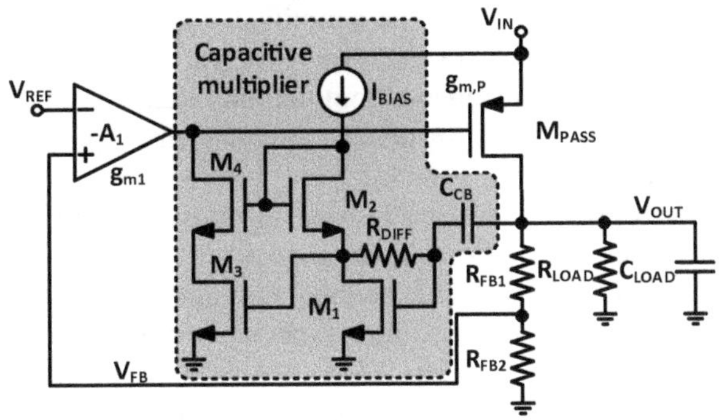

Fig. 2.9 Schematic of the compensation proposed in [37] based on a capacitive multiplier

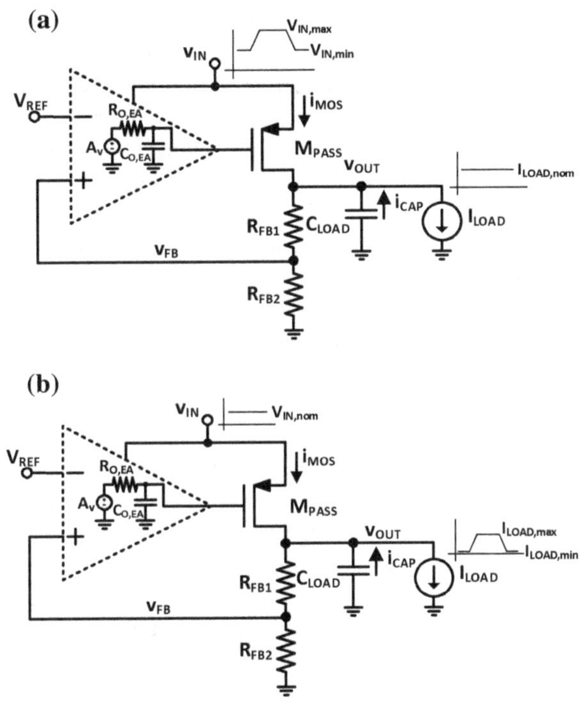

Fig. 2.10 Simplified scheme of an LDO regulator used to analyse its transient response to changes in **a** the line voltage and **b** the load current

2.2.2 *Transient Response*

The effect of variations in the line voltage or current load on the transient response is characterized by the settling time and by the maximum voltage of the spikes that appear at the output. These voltage spikes are undesirable because they can cause an abnormal behaviour in the circuits fed by the regulator. For instance, they might cause the accidental shutdown of a microcontroller, or spurious variations in the frequency of a Voltage Controlled Oscillator (VCO).

Figure 2.10a shows the simplified scheme of an LDO regulator, which will be used to study the transient response to variations in the line voltage. The line voltage is considered to increase from nominal to maximum value, remaining at this value a time long enough to stabilize the transient response. Finally, it returns to its initial value.

Figure 2.11 shows the transient response of the regulator, being t_r and t_f the rise and fall times of the line voltage, respectively. The settling time, t_{set}, is the time elapsed since the application of the input stimulus to the time when the output voltage has entered, and remains within, a specified error band.

In the ideal case (Fig. 2.11a), the response of the control loop is immediate, and thus, there is no change in the source-to-gate voltage, v_{SG}, and current, i_{MOS}, of M_{PASS}, or in the output voltage, v_{OUT}. Figure 2.11b presents a more realistic case, where the control loop has a finite bandwidth. Now, when the line voltage increases, the control circuit takes some time to react, which results in an instantaneous increase of v_{SG} and i_{MOS}. The excess current at the output is diverted to the load capacitor C_{LOAD} causing an undesired increase in v_{OUT}. The process ends when the error amplifier reacts, returning v_{SG} to its initial value, and bringing v_{OUT} to a value close to the nominal value. A dual process occurs when the line voltage drops, returning later to its nominal value. In Fig. 2.11b, the gain of the control loop is assumed to be infinite, so that v_{OUT} returns exactly to its nominal value.

It is also important to study the transient response of an LDO regulator to variations in the load current (Fig. 2.10b). Figure 2.12a shows the transient response of an ideal controller and Fig. 2.12b the more realistic case of a controller with a finite bandwidth control loop.

In the ideal case, when i_{LOAD}, increases, the regulator responds instantaneously, causing an increase of the source-to-gate voltage of M_{PASS}. As a consequence, the current i_{MOS} instantaneously increases to compensate for the increase in the load current. Consequently, there is no excess current at the output, and then, no current flows through C_{LOAD} so that v_{OUT} is maintained at its nominal value.

On the other hand, in Fig. 2.12b, the control loop has a finite bandwidth. When i_{LOAD} increases, v_{SG} does not change instantaneously. The difference between the new load current i_{LOAD} and the current at the output of the regulator has to be provided by C_{LOAD}. As a result, v_{OUT} decreases. After a certain time, the control loop begins to react, bringing v_{SG} to the value that causes M_{PASS} current to equilibrate the increase of i_{LOAD} and, in the last term, to bring v_{OUT} back to its nominal value. It has been assumed, once again, an infinite gain control loop, so that the output voltage recovers exactly its nominal value. A dual process occurs when the load current decreases from its maximum value to its nominal value.

As it can be deduced from this qualitative analysis, the control loop bandwidth is a critical design parameter. Since the dominant pole of an IC-LDO regulator is located at the gate of M_{PASS}, to achieve a good transient response, it will be necessary to design circuits that rapidly charge and discharge the equivalent capacitance at that node. This can be achieved by means of large biasing currents, but would give rise to regulators with poor power efficiency, which is not acceptable in most cases.

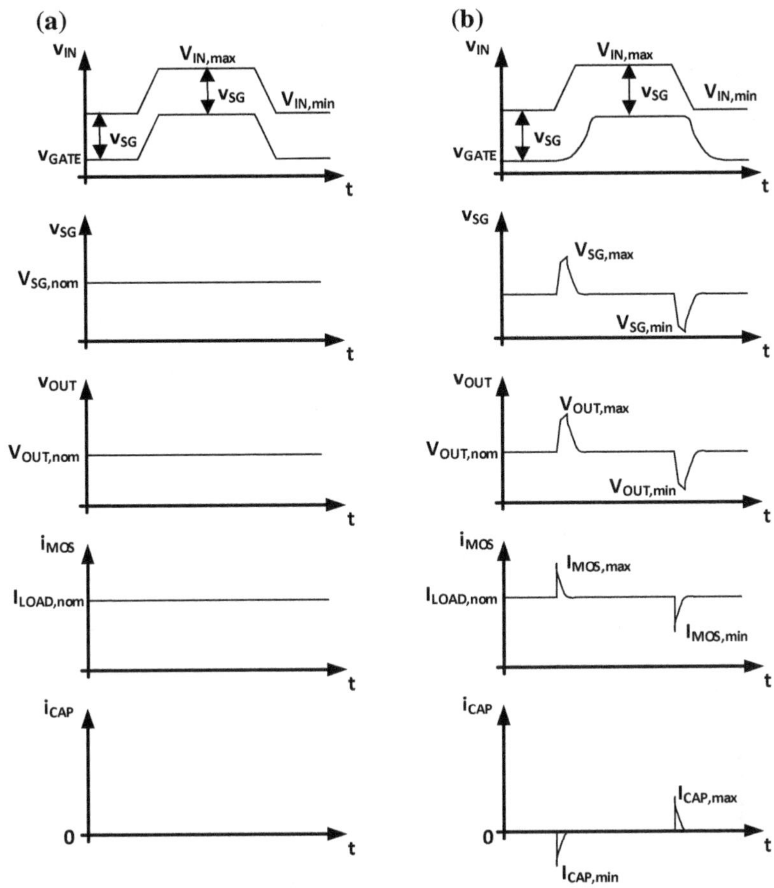

Fig. 2.11 Transient response of an LDO regulator to changes in the input voltage: **a** ideal case and **b** regulator with a limited bandwidth control loop

Most of the solutions proposed to tackle this problem use dynamic biasing circuits that consume a low quiescent current but are capable of providing high transient currents [45–62]. Adaptive biasing, class AB amplifiers and RC couplings are adequate solutions.

One of the first techniques aimed at improving the transient response of an IC-LDO regulator uses adaptive biasing, so that the biasing current is kept small when the load current is small, and it is increased with the changes in I_{LOAD} to improve the speed (and gain) of the control loop. The main disadvantages of this technique lie in (a) the variation of the quiescent consumption with the load current; (b) the complexity of the control block, which must include a circuit to sense I_{LOAD}; and (c) the need to guarantee the stability of the regulator in the entire range of operation.

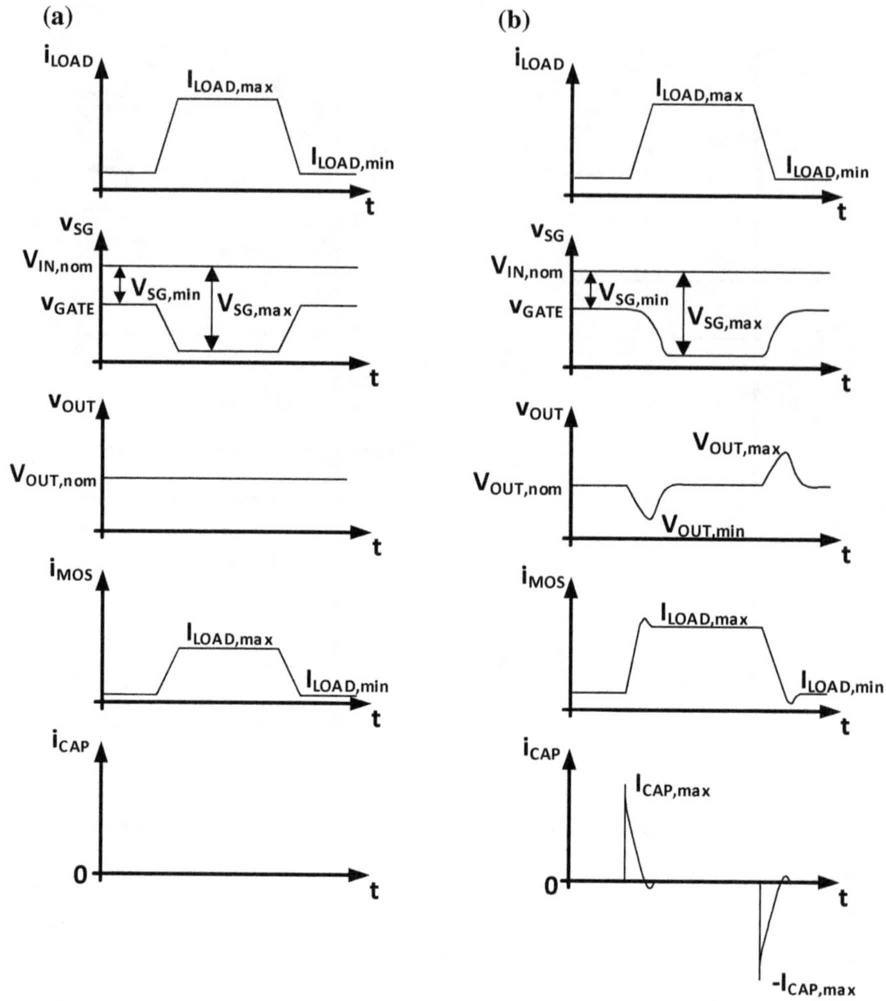

Fig. 2.12 Transient response of an LDO regulator to changes in the load current: **a** ideal case and **b** regulator with a limited bandwidth control loop

In [45], an IC-LDO regulator whose error amplifier makes use of a self-biasing scheme based on [63] is presented. In order to bias the first stage, which is a differential pair, the author implements a current amplifier that converts the error signal at the output of the differential pair in a current. This current charges or discharges the capacitor that sets the bias voltage at the gate of the transistors of the current buffer. As a consequence, the biasing current of the differential pair is now proportional to I_{LOAD}. This causes the open-loop gain of this self-biased operational amplifier to increase by a $g_{m,tail}r_o$ factor, where $g_{m,tail}$ is the transconductance of the transistor that operates as a tail-current source, and r_o, the output resistance of the stage that

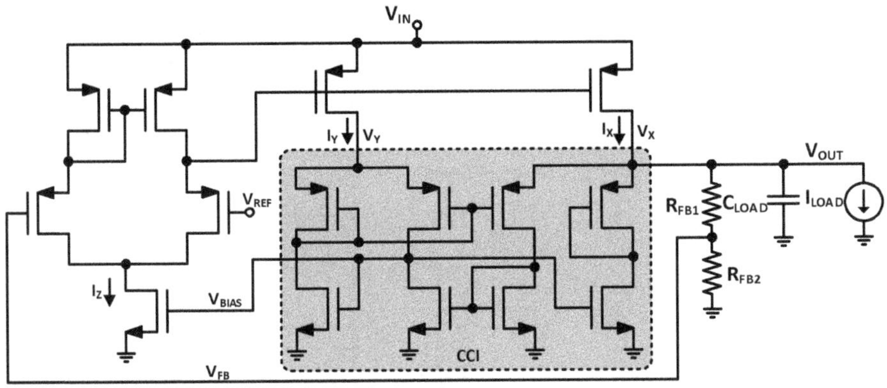

Fig. 2.13 Structure proposed in [46] that makes use of a CCI to sense I_{LOAD} and generates the appropriate bias current I_Z

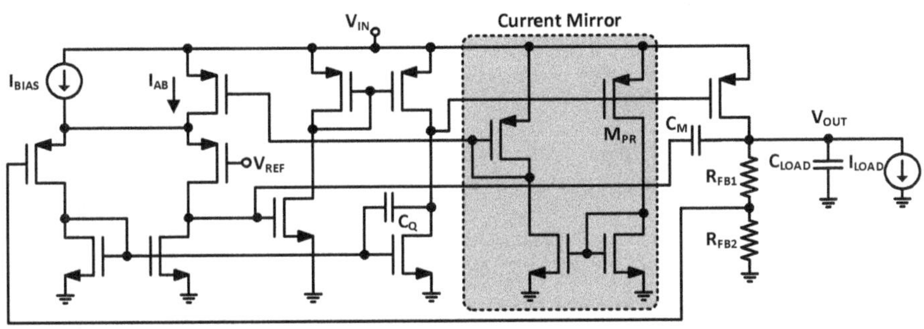

Fig. 2.14 Structure proposed in [48] that makes use of a simple current mirror to replicate the variation of I_{LOAD} in the biasing of the error amplifier

generates the M_{PASS} gate voltage. As a disadvantage, this solution requires a start-up circuit to set an initial charge in the capacitor.

In [46], the authors make use of a modified first-generation current conveyor (CCI) to sense the current delivered to the load and generate a proportional current I_Z that biases the first stage of the error amplifier, Fig. 2.13.

The same idea about sensing the output current is used in references [48] and [52]. In [48], the current source that biases the error amplifier is divided into two. The first one generates a constant current I_{BIAS} of small value, while the second one (I_{AB}) increases with the load current. As shown in Fig. 2.14, a single current mirror generates this current. The size of the transistor that copies the variations in the load current must be fine-tuned because its source-drain voltage, V_{SD}, is larger than the dropout voltage. Ming et al. take advantage of this idea in [53], where a simple current mirror is used to replicate the variations in the load current. Unlike [48], the error amplifier is implemented by the high-bandwidth transconductor presented in [47].

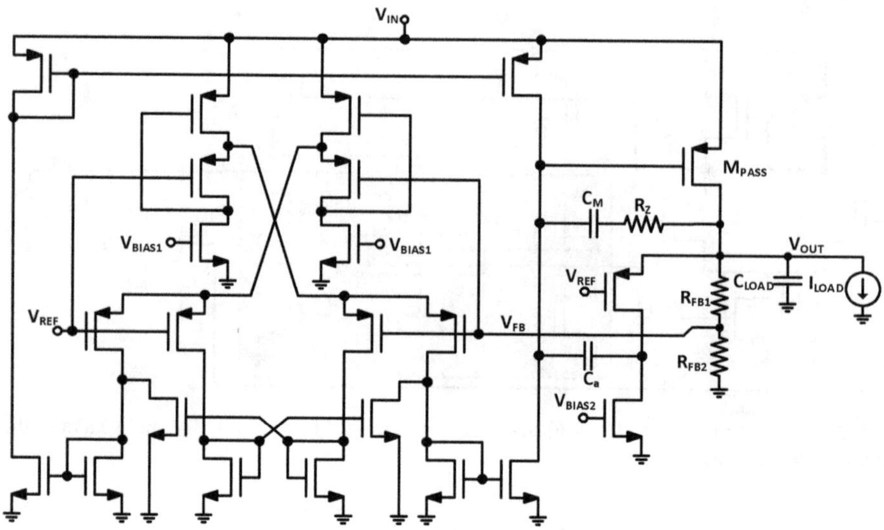

Fig. 2.15 Structure proposed in [56] that makes use of an adaptive biasing scheme for the error amplifier based on the FVF cell with an additional stage for gain improvement

The current conveyor of [46] is used again in [52] to reduce the overshoot at the output node, generating an adaptive biasing for the error amplifier. Like in [46], at a heavy load, the biasing current is large and quickly charges the gate of M_{PASS}, which reduces the overshoot voltage. To reduce the undershoot, which is produced when the biasing current is small, an additional Sub-threshold Undershoot-Reduction (STUR) block is implemented. This circuit differentiates the output voltage so that, when v_{OUT} goes down, it generates a discharging current that helps to reduce the voltage at the gate of M_{PASS}. The transistors in the STUR block are biased in the sub-threshold region, so that they can generate a very high current in the transient regime with a low quiescent current consumption.

In [56], authors propose the use of an adaptive biasing error amplifier, which has high DC gain and slew-rate. This error amplifier is based on the work presented in [64, 65]. The high slew-rate is due to the use of so-called Flipped Voltage Follower (FVF) cell [66, 67], which allows the biasing current of the error amplifier to increase beyond the I_{BIAS} limitation during the transient response. The high DC gain is obtained by the use of two amplifiers in a common-gate configuration whose outputs are added through a current mirror. Figure 2.15 shows the control scheme. The authors include an additional stage to further increase the slew-rate of the error amplifier. This stage contains a common-gate amplifier whose input is V_{OUT}, and whose output is capacitively coupled to the gate of M_{PASS}. The coupling capacitor, C_a, is also used to help in the compensation.

In contrast with the previous works, [54] proposes to divide the pass transistor into two. The first one, of a small size, is responsible for supplying the current to the output when I_{LOAD} is small. The second, of a large size, enters into operation when I_{LOAD}

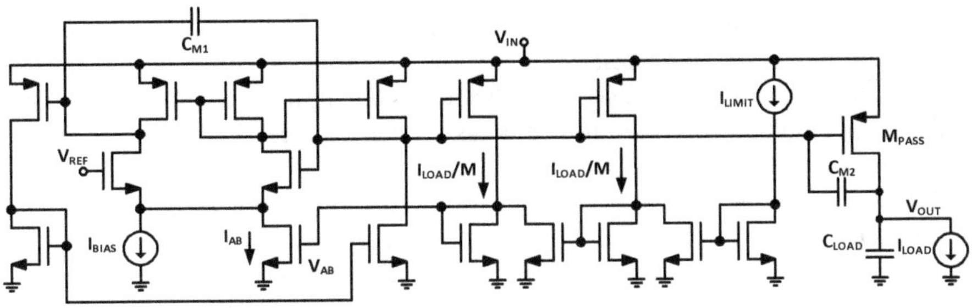

Fig. 2.16 Structure proposed in [62] with adaptive biasing of the error amplifier and NMC

increases. This structure for the pass transistor makes possible to have two control loops, consisting of two and three amplification stages, respectively. Additionally, it increases the stability for small loads, where the regulator has only two poles. Like in [48, 68], the bias current of the error amplifier is dynamically generated. This current is proportional to the load current supplied by the small-sized pass transistor, which is always active, but whose maximum value saturates when the second pass transistor is activated. To overcome this limitation, in [61], the authors present a similar scheme where, unlike [54], I_{LOAD} current is estimated from the gate node voltage.

In [62], an adaptive biasing scheme is proposed for the error amplifier (Fig. 2.16), along with an NMC. Authors study the relationship between the bias current of the error amplifier and the minimum value of the load current for the regulator to be stable. They also optimize the adaptive biasing current of the error amplifier, clamping its value to a critical current when I_{LOAD} is high, so as not to degrade the current efficiency.

Another way to improve the slew-rate of the regulator, without sacrificing the quiescent consumption, is to modify the output stage. In [47], a push–pull output stage is used. Controlled by two common-gate amplifiers and a current summation circuit, it improves the transient response for line and load regulations.

In [49], a circuit is added to the output stage in order to increase the negative slew-rate. To this end, a second fast control loop is added. Unfortunately, the simultaneous operation of the two control loops degrades the stability of the system. In addition, the proposed technique only improves the slew-rate in one direction, so that the quiescent consumption of the error amplifier must remain high if we want to rapidly charge the parasitic capacitance at the gate of M_{PASS} when I_{LOAD} decreases.

In [55], a push–pull output stage is also used. The class AB Operational Transconductance Amplifier (OTA) published in [64] acts as the error amplifier. To improve the slew-rate, an additional circuit is used that increases the charging or discharging currents of the parasitic gate capacitance of M_{PASS} when V_{OUT} is increased. In order to enable this circuit, two complementary current comparators detect voltage spikes at V_{OUT} and activate the corresponding block.

In [57], a push–pull buffer is used to act on the gate voltage of the M_{PASS}. The objective of this buffer is twofold. First, in large signal, it improves the slew-rate of the regulator. Second, in small signal, it decouples the high impedance at the output of the error amplifier and the lumped parasitic capacitance at the gate of M_{PASS}, improving the stability of the system.

Another technique uses RC couplings to improve the slew-rate of the regulator to steep changes in line voltage or load current. An example of this technique is found in [50], where a dynamic biasing circuit is proposed that takes advantage of RC couplings to increase the bias current of the error amplifier current when the output voltage changes. The authors use two comparators to detect changes in V_{OUT} in both directions. In [51], both voltages, V_{OUT} and V_{REF}, are used to increase the current in the output stage of the error amplifier. This mechanism makes LDO regulators suitable to replace switching regulators in systems where power management techniques such as Dynamic Voltage Scale[1] (DVS) are used.

Finally, in [60], the authors modulate the substrate voltage of M_{PASS} to reduce its threshold voltage, allowing a smaller size for the same current. By decreasing the size of the pass transistor, its parasitic gate capacitance decreases, which pushes the dominant pole to a higher frequency and improves the speed of the control loop. In order to modulate the substrate voltage, a second error amplifier is used that fixes an appropriate voltage to prevent the direct biasing of the substrate–source junction. The main drawback of this approach lies in the necessity of guaranteeing a similar behaviour between the main circuit and the replica one in every operating condition.

Tables 2.3 and 2.4 summarize the main performances of the regulators in a set of references cited in this subsection.

2.2.3 *Power Supply Ripple Rejection*

PSRR is a term used to describe the ability of an electronic circuit to suppress changes in the output voltage due to fluctuations in the power supply. Unlike line regulation, the PSRR determines to what extent the low- and/or high-frequency disturbances in V_{IN} will appear in the regulated output voltage. In our case,[2]

$$PSRR = \frac{\Delta v_{OUT}}{\Delta v_{IN}} \tag{2.14}$$

Since the input voltage fluctuations are considered to be of a small value, the PSRR is normally computed in a small-signal analysis.

[1]DVS is a method of reducing the average power consumption in embedded systems. This is accomplished by reducing the switching losses of the system by selectively reducing the frequency and voltage of the system.

[2]Note that this definition is misleading. Despite the PSRR is defined in Eq. 2.14 to be a rejection ratio, it increases when the rejection to power supply perturbations diminishes. However, this definition is maintained here due to its wide use in LDO regulators' literature.

Table 2.3 Comparison of LDO regulators in selected references (1)

		[46]	[47]	[48]	[49]	[50]	[51]	[52]	[53]	[54]
Process	[μm]	0.35	0.18	0.35	0.35	0.13	0.35	0.35	0.35	0.065
V_{IN}	[V]	2.0–3.6	1.0–1.8	1.2–3.3	1.8–3.3	0.9–1.5	1.2–3.3	1.2	2.5–4.0	1.2
V_{OUT}	[V]	1.8	0.9	1.0	1.6–3.1	0.8	1.0–3.1	1.0	2.35	1.0
$V_{DROPOUT}$	[mV]	200	100	200	200	100	200	200	150	200
$I_{LOAD,max}$	[mA]	240	50	100	100	50	100	100	100	100
I_Q[a]	[μA]	1030	1.2	680.5	37.7	1.33	76	380.1	7	82.4
C_{OUT}	[pF]	–[b]	1e2	1e2	1e2	1e4	0	1e2	1e2	1e2
$\eta \vert I_{LOAD,max}$	[%]	99.957	99.998	99.325	99.962	99.999	99.924	99.621	99.993	99.918
Area	$[mm^2]$	0.11[c]	0.9[c,d]	0.099	0.145	0.031	0.350[c]	0.099[c]	0.064	0.017
Response Time[a]	[μs]	0.2[e]	4	0.5[e]	9	140	2.4	5[e]	0.15	6
ΔV_{OUT} varying V_{IN}										
• Maximum	$[mV]$	20	200[e]	60	–[b]	100[e]	175[e]	50	196	8.91
• Minimum	$[mV]$	−20	−120[e]	−70	–[b]	−250[e]	−110[e]	−38	−183	−10.63
$\Delta V_{IN}/t_r^f$	$[V/\mu s]$	0.1/2	0.5/1[e]	0.2/0.1	–[b]	0.6/0.5	1/2	1/1	0.5/0.5	0.2/10
ΔV_{OUT} varying I_{LOAD}										
• Maximum	$[mV]$	10	200[e]	24	97	100[e]	90	50	231	0
• Minimum	$[mV]$	−50	−425[e]	−12	−78	−750[e]	−90	−105	−243	−68.8
$\Delta I_{LOAD}/t_r^f$	$[mA/\mu s]$	149/2	49.50/0.2[e]	25/0.1	100/0.1	50/0.2	99/1	100/1	99.95/0.5	100/0.3
Line Regulation	$[mV/V]$	0.8	3.625	0.22	0.057	–[b]	7.143[e]	0.39	1	4.7
Load Regulation	$[\mu V/mA]$	2.77	148	9.9	109	–[b]	170[e]	78.2	80	300

[a]Worst case, [b]Not available, [c]Effective area, [d]Pads included, [e]Estimation based on published results, [f]t_r: Rise time

Table 2.4 Comparison of LDO regulators in selected references (2)

		[55]	[56]	[57]	[58]	[59]	[60]	[61]	[62]
Process	[μm]	0.35	0.11	0.065	0.065	0.13	0.13	0.13	0.18
V_{IN}	[V]	1.2	1.8–3.8	0.75–1.2	0.75–1.2	0.75–1.2	1.1–1.4	1	1.4
V_{OUT}	[V]	1.0	1.6–3.6	0.5	0.55	0.502	1.0	0.8	1.2
$V_{DROPOUT}$	[mV]	200	200	250	200	148	100	200	200
$I_{LOAD,max}$	[mA]	100	200	50	50	100	5	100	100
I_Q^a	[μA]	14.0	41.5	16.2	487	43.2	99.04	93	141
C_{OUT}	[pF]	1e2	4e1	1e2	1e4	1e2	–[b]	1e2	1e2
$\eta\vert_{I_{LOAD,max}}$	[%]	99.986	99.979	99.968	99.035	99.957	98.058	99.907	99.859
Area	$[mm^2]$	0.038	0.210	0.010[c]	0.013[c]	–[b]	0.025	0.042	0.07
Response Time[a]	[μs]	2.7	0.65	1.2	20[d]	12.3	1[d]	1.7	30[d]
ΔV_{OUT} varying V_{IN}									
• Maximum	$[mV]$	–[b]	–[b]	36.25	4.1	–[b]	4[d]	–[b]	–[b]
• Minimum	$[mV]$	–[b]	–[b]	−41.88	−4.3	–[b]	−4[d]	–[b]	–[b]
$\Delta V_{IN}/t_r^e$	$[V/\mu s]$	–[b]	–[b]	0.45/5	0.45/10	–[b]	0.27/0.1[d]	–[b]	–[b]
ΔV_{OUT} varying I_{LOAD}									
• Maximum	$[mV]$	200[d]	200	96.9	29	92	200[d]	90	85
• Minimum	$[mV]$	−270	−385	−100	−113	−270	−220[d]	−120	−110
$\Delta I_{LOAD}/t_r^e$	$[mA/\mu s]$	99.9/1	199.5/0.5	50/0.1	50/0.1	100/0.1	1/0.1[d]	99.999/0.1	99.99/1
Line Regulation	$[mV/V]$	–[b]	8.9	6.67	4	2.86	54.48[d]	1	0.6
Load Regulation	$[\mu V/mA]$	–[b]	108	560	180	–[b]	–[b]	140	270

[a] Worst case, [b] Not available, [c] Effective area, [d] Estimation based on published results, [e] t_r: Rise time

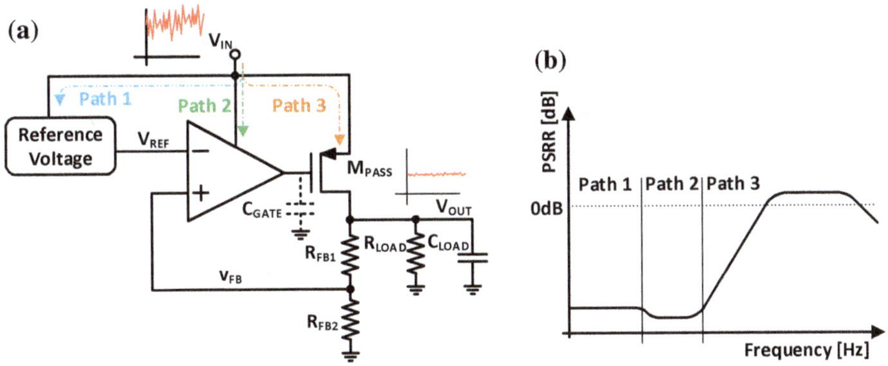

Fig. 2.17 Contributions to the PSRR in an LDO regulator: **a** signal paths and **b** typical spectrum of the PSRR showing the dominant contribution in each frequency range

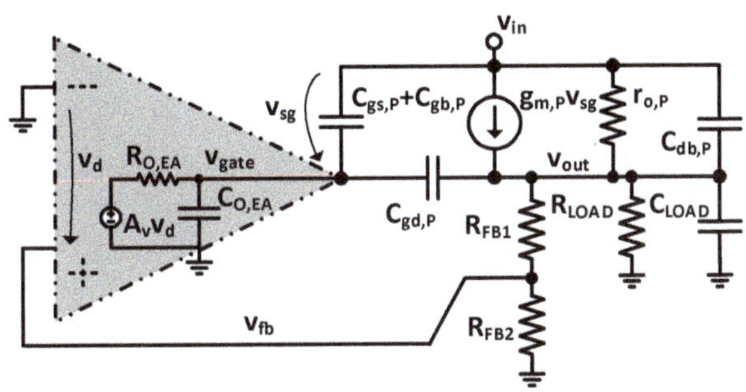

Fig. 2.18 Small-signal model used for PSRR analysis

Figure 2.17a shows the different paths that affect the PSRR in an LDO regulator. Figure 2.17b shows the typical frequency response of the PSRR, indicating which path usually dominates in each frequency range. The first path comes from the circuit that generates the reference voltage, V_{REF}, and normally dominates the PSRR in low frequency. The second one comes from the error amplifier and usually dominates at intermediate frequencies. The latter has two contributions: the first one comes from the control loop, and the second one comes directly from M_{PASS} through its finite output resistance and its parasitic capacitances, which couple the line to the output node at high frequencies.

To study the different contributions to the PSRR, the simplified small-signal model of Fig. 2.18 will be used, where the error amplifier is represented by a gain block, Av, with a single pole located at $\omega_{p1} = \frac{1}{R_{o,EA}C_{GATE}}$. As usual, C_{GATE} is the total capacitance at the gate of M_{PASS}. A small signal model, valid for intermediate frequencies, has been chosen for M_{PASS}. Note that, for EC-LDO regulators, it would be necessary to take also into account the compensating capacitor at the output node, including its parasitic inductance and equivalent series resistance, which is beyond the interest of this study.

Solving the nodal equations, Eq. 2.15 is obtained:

$$PSRR(s) = \frac{V_{out}(s)}{V_{in}(s)} = \frac{R_{OUT}}{r_{o,P}} \frac{1 + g_{m,P}r_{o,P} + s \cdot a_1 + s^2 \cdot a_2}{1 + g_{m,P}R_{OUT}A_v\beta + s \cdot b_1 + s^2 \cdot b_2} \tag{2.15}$$

where

$$R_{OUT} = r_{o,P} \| (R_{FB1} + R_{FB2}) \| R_{LOAD} \tag{2.16}$$

$$\beta = \frac{R_{FB2}}{R_{FB1} + R_{FB2}} \tag{2.17}$$

$$a_1 = R_{O,EA}(1 + g_{m,P}r_{o,P})\left(C_{GATE} + C_{gd,P}\right) + r_{o,P}\left(C_{db,P}\right) \tag{2.18}$$

$$a_2 = r_{o,P}R_{O,EA}\left(C_{db,P}\right)\left(C_{GATE} + C_{gs,P} + C_{gd,P}\right) + r_{o,P}R_{O,EA}C_{gs,P}C_{gd,P} \tag{2.19}$$

$$b_1 = R_{OUT}\left(C_{LOAD} + C_{gd,P} + C_{db,P}\right) + R_{O,EA}\left(C_{GATE} + C_{gs,P} + C_{gd,P}\right) + + g_{m,P}R_{O,EA}C_{gd,P}R_{OUT} - R_{OUT}C_{gd,P}A_v\beta \tag{2.20}$$

$$b_2 = R_{O,EA}R_{OUT}\left[\left(C_{LOAD} + C_{gd,P} + C_{db,P}\right)\left(C_{GATE} + C_{gs,P} + C_{gd,P}\right) - -C_{gd,P}^2\right] \tag{2.21}$$

Assuming, as usual, that $g_{m,P}r_{o,P} \gg 1$, and particularizing Eq. 2.15 for low frequencies (i.e. $s \rightarrow 0$),

$$PSRR(0) = \frac{R_{OUT}}{r_{o,P}} \frac{1 + g_{m,P}r_{o,P}}{1 + g_{m,P}R_{OUT}A_v\beta} \approx \frac{1}{A_v\beta} \tag{2.22}$$

This expression matches that obtained for line regulation in Eq. 2.6. This result was expected since the line regulation measures the ability to maintain the regulated output voltage under steady-state variations of V_{IN}.

Equations 2.15–2.21 model the PSRR for any LDO regulator, regardless the value of C_{LOAD}. To estimate the contribution of each parasitic capacitance of M_{PASS} and, hence, determinate which ones can be neglected, some simulations have been carried out. Concretely, the circuits shown in Fig. 2.19 have been simulated in Spectre® Circuit Simulator using the circuit parameters of Table 2.5 in a standard 65 nm CMOS technology. The results for each model can be observed in Fig. 2.20a, b, assuming a C_{LOAD} in the order of picofarads or microfarads, respectively. Figure 2.20a shows the typical PSRR curve for an IC-LDO regulator.

It can be observed that the effect of $C_{db,P}$ is critical at high frequency. Hence, this parasitic capacitance cannot be neglected. In addition, for frequencies greater than $\omega_{z1} = \frac{1}{R_{O,EA}\left(C_{GATE}+C_{gd,P}\right)}$, the PSRR worsens, as a consequence of the decrease of the gain in the regulation loop. This worsening persists until the effect of the output impedance becomes dominant.

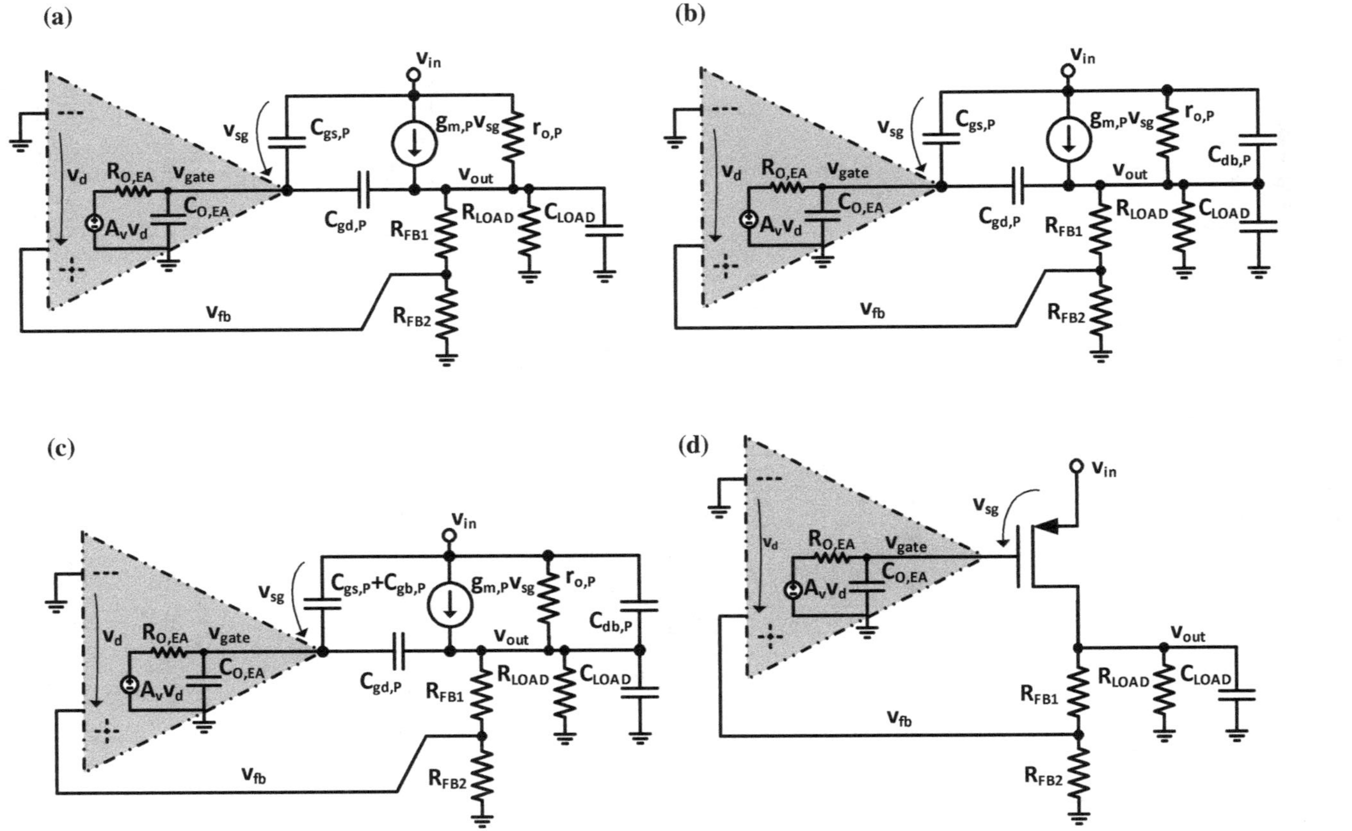

Fig. 2.19 Models used to simulate the PSRR response: **a** without $C_{db,P}$ or $C_{gb,P}$, **b** with $C_{db,P}$, without $C_{gb,P}$, **c** with $C_{db,P}$ and $C_{gb,P}$ and **d** M_{PASS} model provided by the Process Design Kit

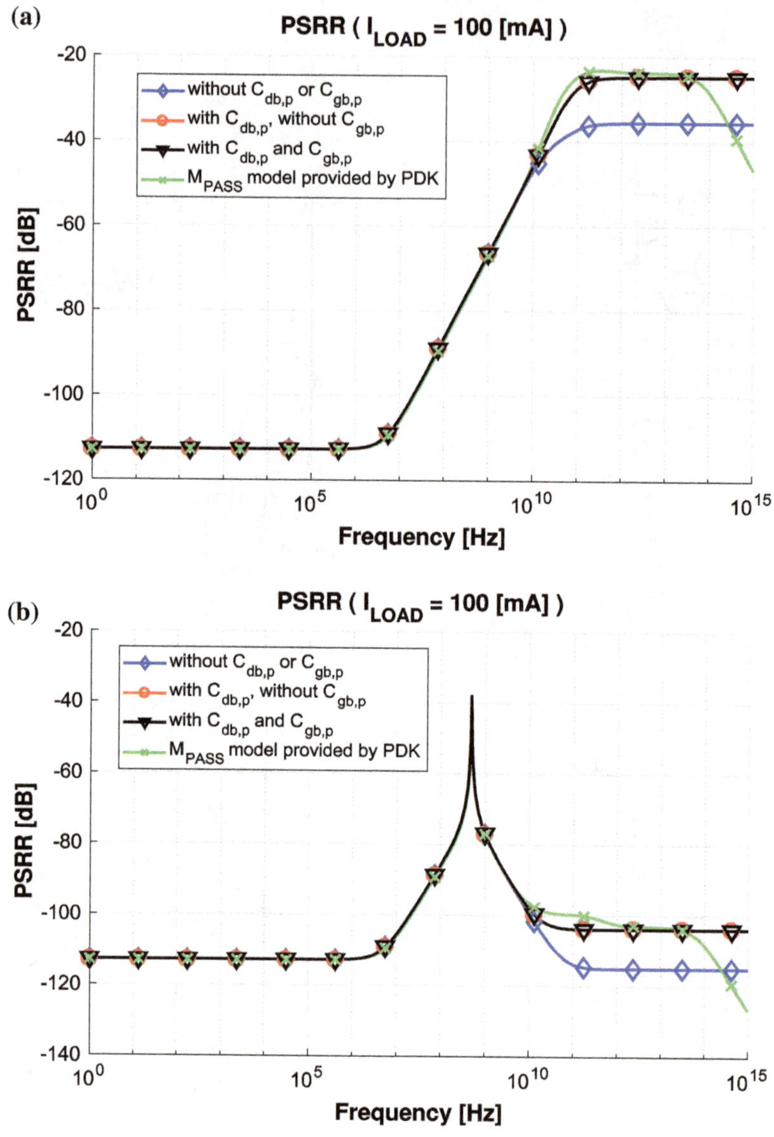

Fig. 2.20 Spectrum of the PSRR for: **a** internally compensated LDO regulator (C_{LOAD} = 100 pF), and **b** externally compensated LDO regulator (C_{LOAD} = 1 μF), using the models shown in Fig. 2.19

Neglecting the capacitor ESR and parasitic capacitances. These assumptions can be also used for EC-LDO regulators (Fig. 2.20b). In this case, the output capacitor where the output capacitor has a large value, and the effect of the output impedance is dominant at medium and high frequencies. Similarly, as for IC-LDO regulators, the effect of $C_{db,P}$ cannot be neglected.

Table 2.5 Parameter values used to obtain the curves of Fig. 2.20

Magnitude	Value	Magnitude	Value
A_V	10^6 [V/V]	$g_{m,P}$	0.776 [Ω^{-1}]
R_{FB1}	10 [kΩ]	$r_{o,P}$	8.375 [Ω]
R_{FB2}	10 [kΩ]	$C_{gs,P}$	10.85 [pF]
R_{LOAD}	10 [Ω]	$C_{gd,P}$	6.685 [pF]
$R_{O,EA}$	1 [kΩ]	$C_{gb,P}$	0.779 [pF]
C_{GATE}	25 [pF]	$C_{db,P}$	4.636 [pF]
W/L	250×(50/0.18) [μm]/[μm]		

From the analysis above, it can be concluded that IC-LDO regulators generally presents a bad PSRR for medium and high frequencies. This problem has been addressed in references [69–85].

One of the simplest approaches to improve the PSRR at medium and high frequencies introduces a low-pass RC filter in series with the output terminal that attenuates the high-frequency variations at the output voltage. However, to minimize the dissipated power and the dropout voltage, the filter resistance must be very low (in the range of 1–10 Ω) and thus, a very large capacitor (in the range of 1–100 μF) is required. As a consequence, this technique is not valid in applications requiring a large scale of integration, low supply voltage and low power consumption, typical of SoCs. Even though an RLC implementation would allow a reduction in the capacitor size, the large value of the inductance and the power that would be dissipated in its parasitic resistance make this implementation equally unfeasible.

In [70, 71], a technique for the improvement of the PSRR is proposed based on the use of an NMOS transistor that acts as a cascode for M_{PASS} [86]. In this way, this transistor is isolated from the fluctuations of the input voltage. Figure 2.21 shows a simplified scheme of the proposed circuit. Authors report an improvement of the PSRR of about 30 dB for medium and high frequencies. However, this technique shows several disadvantages. For a proper operation of the cascode transistor, its gate voltage must be higher than V_{IN}. The authors generate this voltage by means of a charge pump, whose output is filtered to minimize noise. Another drawback is the increase in the dropout voltage (0.6 V), inappropriate for current technologies. Not to mention the corresponding loss of efficiency, and the increase in area due to the cascode transistor (which carries the same current as M_{PASS}), the charge pump and the low-pass filter.

In [72, 73], a similar technique was used. The reference circuit is replaced by a linear regulator with an NMOS pass transistor. It allows to reduce the power dissipation with respect to reference [71], as well as the size of the capacitors of the charge pump. In [73, 74], additional regulators are used to increase the immunity of the gate voltage of the cascode transistor, at the expense of an even larger dropout voltage.

A second negative feedback loop to improve the PSRR in medium and high frequencies using a replica circuit is added in [75]. To simplify the implementation,

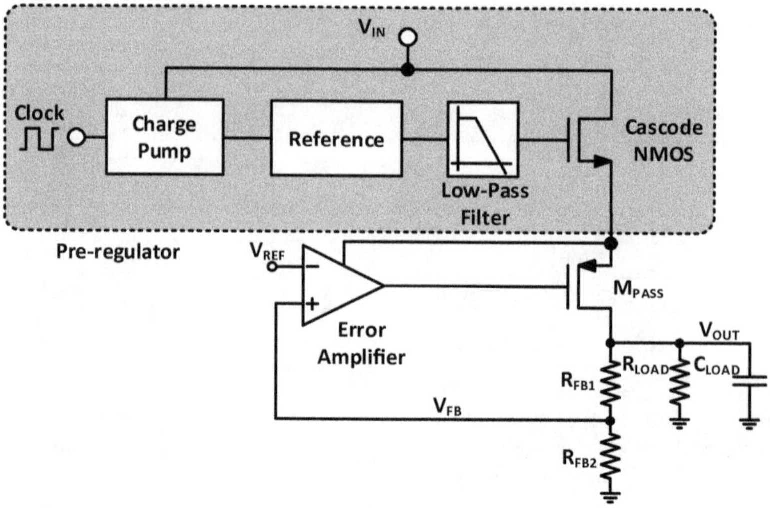

Fig. 2.21 Simplified block diagram of the PSRR reduction technique in [70, 71] where a cascode transistor is inserted in series with M_{PASS}

both feedback loops are merged in a single current amplifier. The main drawback of this approach lies in the necessity of guaranteeing a similar behaviour between the main circuit and the replica one in every operating condition.

The most effective way to improve the PSRR is to implement a direct path between the input signal V_{IN}, connected to the source of the pass transistor, and its gate (Fig. 2.22). In this way, the voltage V_{SG} of the transistor (and, in a first-order approach, its current) is independent on V_{IN}. Different implementations of this technique have been proposed. In [76, 77], the effect of the output resistance $g_{ds,p}$ of M_{PASS} was taken into account, and the direct fast path amplifies V_{IN} variations by the ratio $\frac{g_{m,P}+g_{ds,p}}{g_{m,P}}$. Also, to avoid interfering with the main control loop, it is convenient to generate a zero in the PSRR transfer function to cancel the effect of the LDO regulator's dominant pole. Using this technique, the authors extended the small value of the PSRR, typical of low frequencies, up to 9 MHz.

A similar technique was presented in [80], where a band–band filter was introduced as a direct path to couple the input voltage variations to the output. The second stage of the error amplifier is used to make the summation of the direct path current with that of the main loop. This technique only improves the PSRR at intermediate frequencies, requiring the use of additional techniques to enhance the PSRR at low and high frequencies.

In [83], a current proportional to the variations of V_{IN} is generated by means of a scaled replica of M_{PASS} whose current is correlated to I_{LOAD}, so that the small-signal parameters of the replica transistor have values similar to those of M_{PASS}. At the gate of the replica transistor, the input signal fluctuations are coupled through a diode connection. This allows to generate a current proportional to those fluctuations.

Table 2.6 Comparison of selected references addressing a good PSRR

		[71]	[72]	[73]	[74][a]	[74][b]	[78]	[79]	[80]	[83]
Process	[μm]	0.6	0.35	0.35	0.35	0.35	1.5	0.18	0.13	0.18
V_{IN}	[V]	1.8	1.6	1.8	1.6	1.8	1.4	1.8	1.15–1.4	1.8
V_{OUT}	[V]	1.2	1.2	1.2	1.2	1.2	1.2	1.5	1.0	1.6
$V_{DROPOUT}$	[mV]	400	150	200	400	600	200	300	150	200
I_{LOAD}	[mA]	5	10	10	12	12	5	25	50	50
I_Q[c]	[μA][c]	70	70	39	28.6	43.9	192[d]	300	37	55
C_{LOAD}	[pF]	–[e]	1e2	1e2	1e2	1e2	1.5e4	1.25e2	2e1	1e2
$\eta\|_{I_{LOAD,max}}$	[%]	98.619	99.305	99.612	99.762	99.636	96.300	98.814	99.926	98.912
Area	[mm^2]	–[e]	0.066[f]	0.092[f]	0.055	0.084	1.2	0.04	0.018	0.14
Response Time[c]	[μs]	0.6[d]	2	3[d]	10	10	800	–[e]	0.4	6
ΔV_{OUT} varying V_{IN}										
• Maximum	[mV]	–[e]	–[e]	–[e]	61	8	50[d]	–[e]	–[e]	5.5[d]
• Minimum	[mV]	–[e]	–[e]	–[e]	−62	−8	15–[d]	–[e]	–[e]	−7.5[d]
$\Delta V_{IN}/t_r^g$	[$V/\mu s$]	–[e]	–[e]	–[e]	1.7/0.5	1.7/0.5	0.5/1	–[e]	–[e]	0.8/10
ΔV_{OUT} varying I_{LOAD}										
• Maximum	[mV]	290[d]	303	360	40	42	170[d]	–[e]	56	120
• Minimum	[mV]	−750[d]	−410	−380	−100	−105	−60[d]	–[e]	−42	−80
$\Delta I_{LOAD}/t_r^g$	[$mA/\mu s$]	5/0.1[d]	10/0.06	5/0.01	12/1	12/1	5/0.1	–[e]	49.95/0.2	50/0.1
Line Regulation	[mV/V]	–[e]	0.25	0.45	0.31	0.28	–[e]	–[e]	8.1	–[e]
Load Regulation	[$\mu V/mA$]	–[e]	320	500	390	680	920	–[e]	55.6	140
PSRR @f=1[kHz]	[dB]	−70[d]	−67[d]	–[e]	−70[d]	−82[d]	–[e]	–[e]	−72[d]	–[e]
PSRR @f=100[kHz]	[dB]	−62[d]	−39[d]	−55[d]	−35[d]	−60[d]	−57	–[e]	−40[d]	−60[d]
PSRR @f=1[MHz]	[dB]	−40[d]	−25[d]	−42[d]	−25[d]	−42[d]	−55	−40	−40	−70

[a]One pre-regulator, [b]Two pre-regulators, [c]Worst case, [d]Estimation based on published results, [e]Not available, [f]Effective area, [g]t_r: Rise time

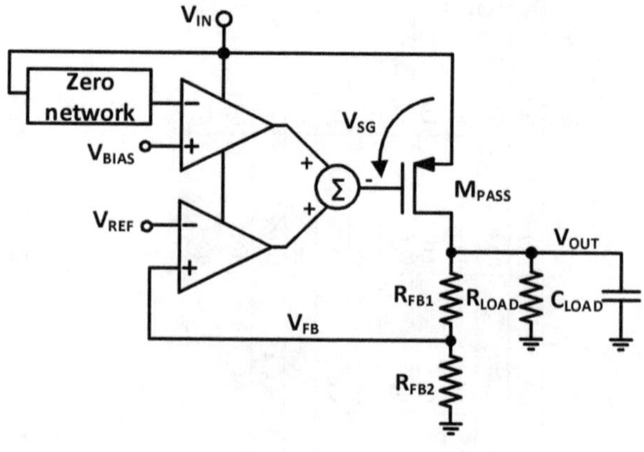

Fig. 2.22 Block diagram of the PSRR improvement technique proposed in [77]

This current is amplified to ensure that the transfer function between V_{IN} and the gate voltage of the replica transistor is unitary. Then, it is converted to voltage and added to the gate of M_{PASS}. With this process, the voltage V_{SG} of M_{PASS} is independent on the fluctuations of V_{IN}, at least up to the frequency at which the circuit introduced for PSRR improvement stops working.

As a summary of this section, Table 2.6 shows a comparison of the selected references that address the PSRR limitations of IC-LDO regulators.

2.3 Comparison of the State of the Art

To carry out a fair comparison between existing IC-LDO regulators, some Figures of Merit (FOMs) have been proposed in the literature.

To this end, we select the FOM proposed in [87], which takes into account the quiescent current consumption (I_Q), the maximum load current ($I_{LOAD,max}$) and the response time (T_r), which is defined as

$$T_r = \frac{C_{LOAD} \Delta v_{OUT}}{I_{LOAD,max}} \tag{2.23}$$

where Δv_{OUT} is the maximum variation of the voltage at the regulator output when there is a change in the line voltage or the load current. Equation 2.24 shows the expression for FOM_1. Note that the smaller the FOM_1, the better the regulator.

$$FOM_1 = T_r \frac{I_Q}{I_{LOAD,max}} = \frac{C_{LOAD} \Delta v_{OUT}}{I_{LOAD,max}} \frac{I_Q}{I_{LOAD,max}} = C_{LOAD} \frac{\Delta v_{OUT} I_Q}{I^2_{LOAD,max}} \tag{2.24}$$

Table 2.7 FOM_1 values for selected references

		[30]	[37]	[40][a]	[41][a]	[42]	[44]	[46]	[47]
T_r	[ns]	0.19	0.32	42	0.36	2.8	95	0.025	1.25
FOM_1	[fs]	72.2	416	16380	388.8	2520	42750	107.29	30
		[48]	[49]	[50]	[52]	[53]	[54]	[55]	[56]
T_r	[ns]	0.036	0.175	170	0.155	0.474	0.069	0.47	0.117
FOM_1	[fs]	245	65.98	4522	589.16	33.18	56.69	65.8	24.28
		[57]	[58]	[59]	[61]	[62]	[72]	[73]	[74][b]
T_r	[ns]	0.394	28.4	0.362	0.21	0.195	7.13	7.4	1.167
FOM_1	[fs]	127.59	2.77e5	156.38	195.3	274.95	4.99e4	2.89e4	2.78e3
		[74][c]	[78]	[80]	[83]				
T_r	[ns]	1.225	690	0.039	0.4				
FOM_1	[fs]	4.48e3	2.65e7	29.01	440				

[a]Simulation results, [b]One pre-regulator, [c]Two pre-regulators

Table 2.7 shows the value of FOM_1 for a selected set of regulators. Figure 2.23 represents the value of FOM_1 versus quiescent current consumption. In this case, the closer to the origin, the better the regulator. The selected regulators are represented in this figure with a different colour, depending on their main design objective. Those that target good stability are coloured in blue, while those that target a good transient response and a good PSRR are depicted in red and black, respectively. It is important to highlight that this classification is subjective, and some of the selected papers target more than one performance at the same time.

According to the results of Table 2.7, the controller that obtains the best FOM_1, among those selected for comparison, is that of reference [56], with a value of 24.28 fs. This work aims at improving the transient response using an amplifier with adaptive biasing based on references [64, 65]. Despite the low value of FOM_1, Fig. 2.23 shows that its I_Q is not among the smallest ones, as it significantly increases with the load current.

The second best regulator is that of the reference [80]. Although the main objective of its authors was an improvement of the PSRR, they achieve a good transient response thanks to the error amplifier of reference [33]. It has a push–pull output stage where the PMOS transistor is controlled by the output of the first stage, while the NMOS transistor is controlled by the negative output of the input differential pair. Like reference [56], it achieves a very low value of FOM_1, even though it does not exhibit a low current consumption.

Among the references with the best FOM_1, it is remarkable that the work published in [47] achieves the lowest static consumption of the selected references (1.2 μA).

It is important to highlight that the FOM_1 favours those regulators performing output voltage spikes and, in second term, those with a low quiescent current consumption. Even though some other figures of merit have been proposed that favour other features of the regulator, FOM_1 is still the preferred figure of merit in the literature.

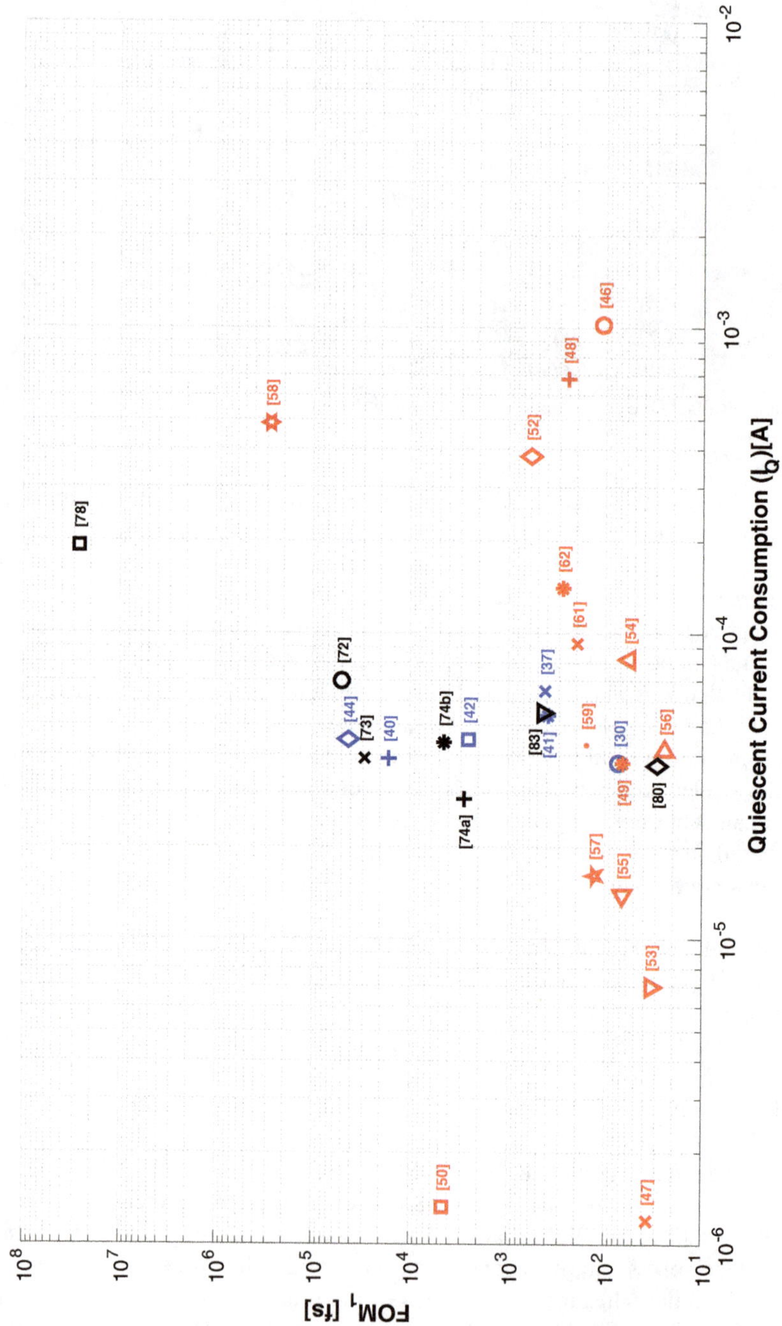

Fig. 2.23 FOM_1 versus quiescent current consumption

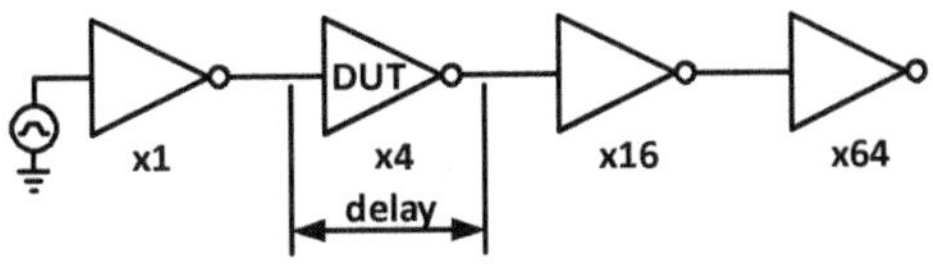

Fig. 2.24 Circuit used to estimate the delay parameter $FO_{4,delay}$, based on a chain of inverters

Reference [47] proposes a new figure of merit that replaces T_r of Eq. 2.23 by the measured settling time, resulting in Eq. 2.25.

$$FOM_2 = t_{set}\frac{I_Q}{I_{LOAD,max}} \tag{2.25}$$

Even though the spikes at the output voltage are highly dependent on the rise, t_r, and fall, t_f, times of the stimulus signal, neither FOM_1 nor FOM_2 defines a standard value for them. In [88], a new FOM (FOM_3) is defined that normalizes the effect of t_r and t_f on the transient response. FOM_3 assumes $t_r = t_f$, and uses a normalization parameter to ensure a fair comparison. This parameter K is given by $K = \frac{t_r}{t_{r,min}}$, where $t_{r,min}$ is the minimum rise time employed in the set of references selected for comparison.

$$FOM_3 = K\frac{\Delta v_{OUT} I_Q}{\Delta I_{LOAD}} \tag{2.26}$$

In [89], two new figures of merit are proposed. They are defined by Eqs. 2.27 and 2.28, respectively. In the first one, FOM_4, the term K in Eq. 2.26 is affected by the exponent 1/3, since the dependence between ΔV_{OUT} and $t_f = t_r$ is not linear. At the same time, this FOM tries to be independent on the technology. To this end, a new delay parameter, $FO_{4,delay}$, is defined, as the delay introduced by a unitary inverter when it is placed at the circuit of Fig. 2.24 [90–92]. It can be observed that the excitation and the load are implemented by means of inverters of the same technology, of relative size 1/4 and 4, respectively. The second FOM proposed in [89] (FOM_5) eliminates the factor K used in FOM_3 and FOM_4, but maintains the independence with a technology given by the $FO_{4,delay}$ parameter. Note that FOM_5 is dimensionless. Additionally, FOM_5, like FOM_1, uses the response time T_r defined in Eq. 2.23.

$$FOM_4 = K^{1/3}\frac{\Delta v_{OUT}\left[I_Q + I_{LOAD}|^{min}\right]}{FO_{4,delay}\Delta I_{LOAD}} \tag{2.27}$$

$$FOM_5 = T_r\frac{I_Q + I_{LOAD}|^{min}}{FO_{4,delay}\Delta I_{LOAD}} \tag{2.28}$$

Table 2.8 Values of the FOMs for the regulators in selected references

		[30]	[37]	[40][a]	[41][a]	[42]	[44]
FOM_1	[fs]	72.2	416	16380	388.8	2520	42750
FOM_2	[ps]	760	19500	9750	2160	2250	1800
FOM_3	[mV]	7.22	10.4	16.38	19.44	23.31	6.3
FOM_4	$[V/\mu s]$	1.68	9.23	6.50	11.94	7.36	1.42
FOM_5	$[10^{-12}]$	1.905	7.83	3.33	14.29	3.06	2.19
		[46]	[47]	[48]	[49]	[50]	[52]
FOM_1	[fs]	107.29	30	245	65.98	4522	589.16
FOM_2	[ps]	858.3	96	3402.5	3393	3640	9120
FOM_3	[mV]	51.5	0.72	2.45	0.66	0.44	58.9
FOM_4	$[V/\mu s]$	10.24	34.78	3.86	0.97	1.099	19.089
FOM_5	$[10^{-12}]$	58.39	5.42	4.97	0.26	0.095	26.53
		[53]	[54]	[55]	[56]	[57]	[58]
FOM_1	[fs]	33.18	56.69	65.8	24.28	127.59	2.77e5
FOM_2	[ps]	10.5	4944	378	134.88	388.8	1.948e5
FOM_3	[mV]	1.66	1.7e6	6.58	6.07	0.64	13.83
FOM_4	$[V/\mu s]$	6.77	645.25	16.94	126.62	5.04	109.15
FOM_5	$[10^{-12}]$	1.940	9.06e6	7.76	29.38	1.19	35.68
		[59]	[61]	[62]	[72]	[73]	[74][b]
FOM_1	[fs]	156.38	195.3	274.95	4.99e4	2.89e4	2.78e3
FOM_2	[ps]	5313.6	1581	42300	14000	11700	23833.3
FOM_3	[mV]	1.56	15.62	27.50	29.95	2.89	16.68
FOM_4	$[V/\mu s]$	6.17	15.58	18.08	61.696	45.03	8.36
FOM_5	$[10^{-12}]$	0.791	13.77	19.98	2.86	0.61	8.11
		[74][c]	[78]	[80]	[83]		
FOM_1	[fs]	4.48e3	2.65e7	29.01	440		
FOM_2	[ps]	36583.3	3.07e7	296	6600		
FOM_3	[mV]	26.89	8.83e7	1.45	2.2		
FOM_4	$[V/\mu s]$	13.47	3020.31	8.49	6.27		
FOM_5	$[10^{-12}]$	12.44	6.10e6	6.38	1.46		

[a]Simulation results, [b]One pre-regulator, [c]Two pre-regulators

Table 2.8 shows the values of the different FOMs defined above for a selected set of references. Figures 2.23, 2.25, 2.26, 2.27 and 2.28 depict these FOMs against the quiescent current of the regulator, maintaining the same convention about colours used in Fig. 2.23.

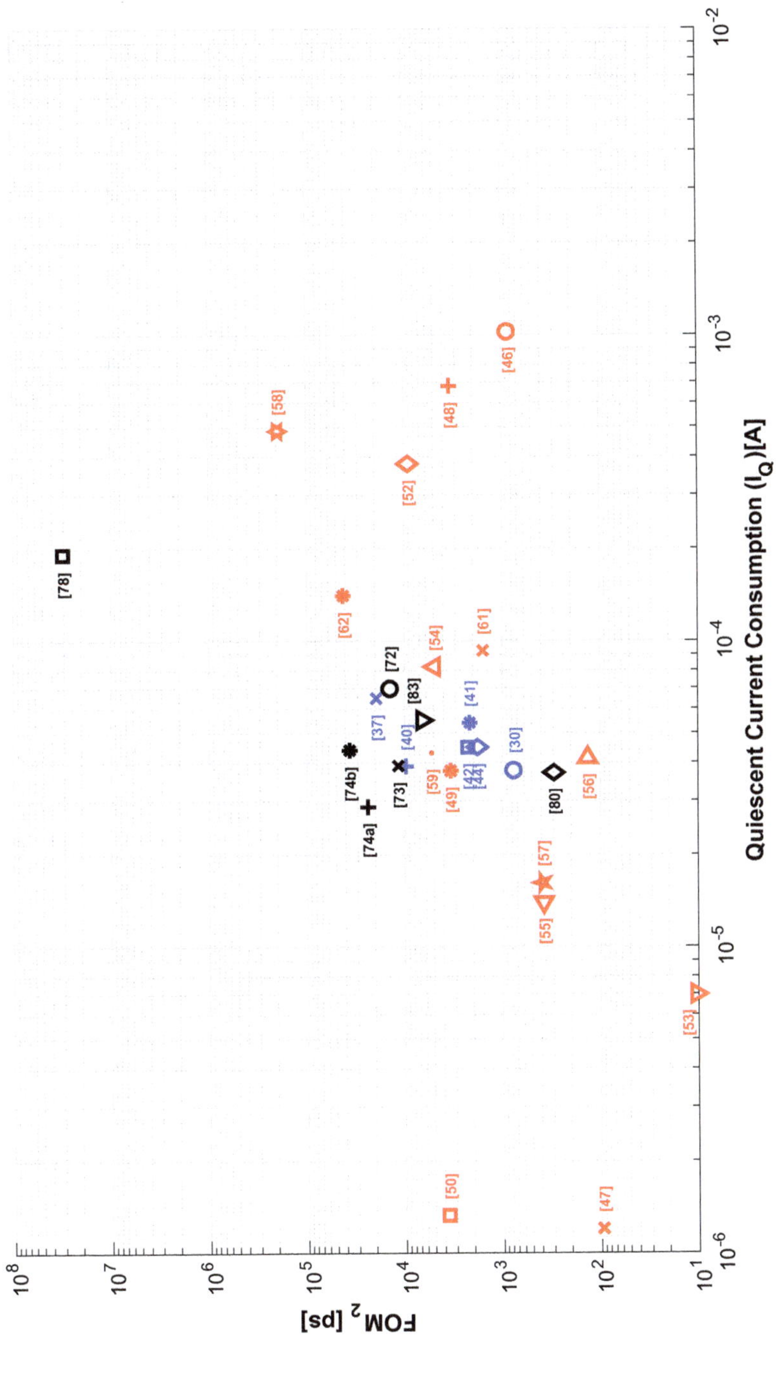

Fig. 2.25 FOM_2 versus quiescent current consumption

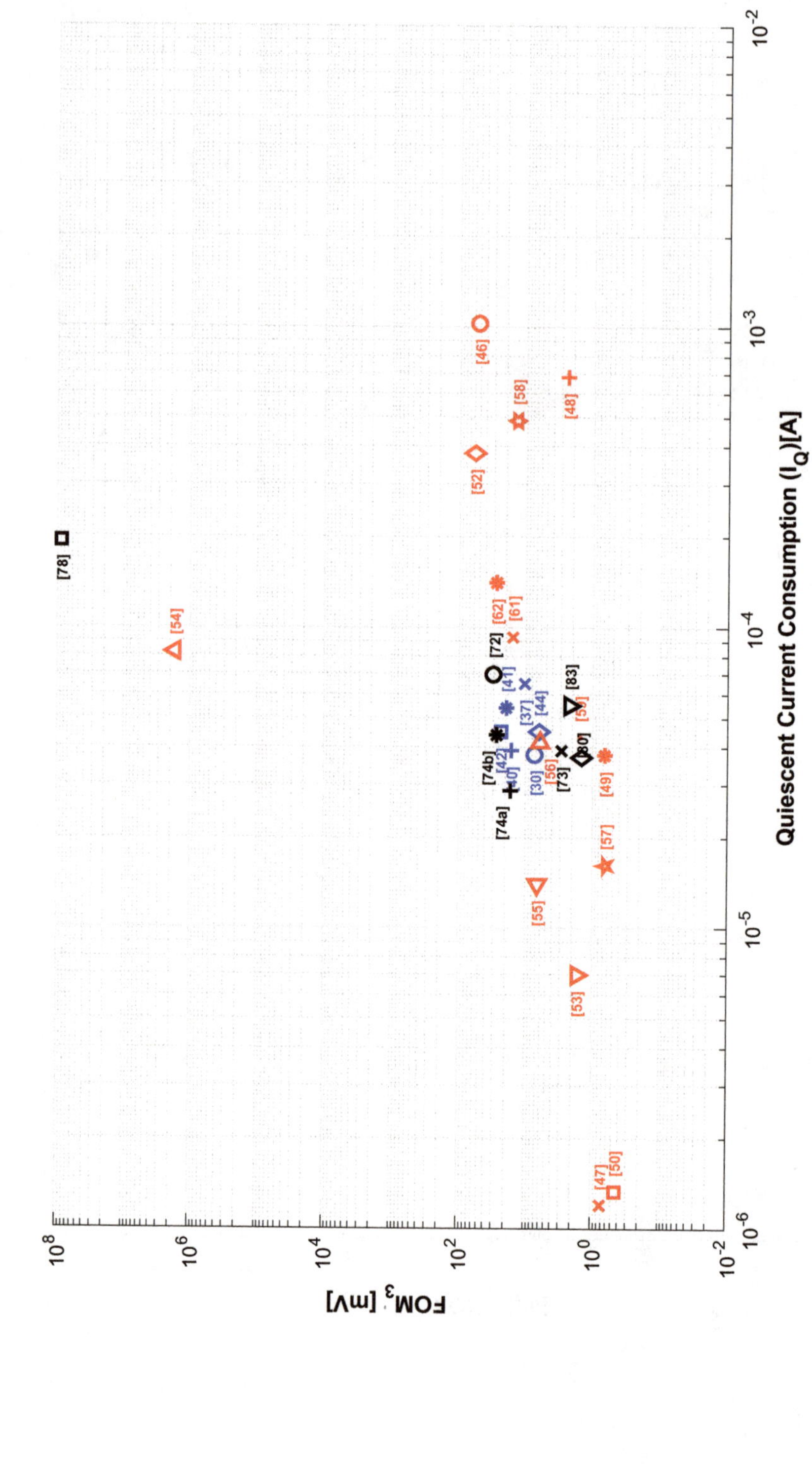

Fig. 2.26 FOM_3 versus quiescent current consumption

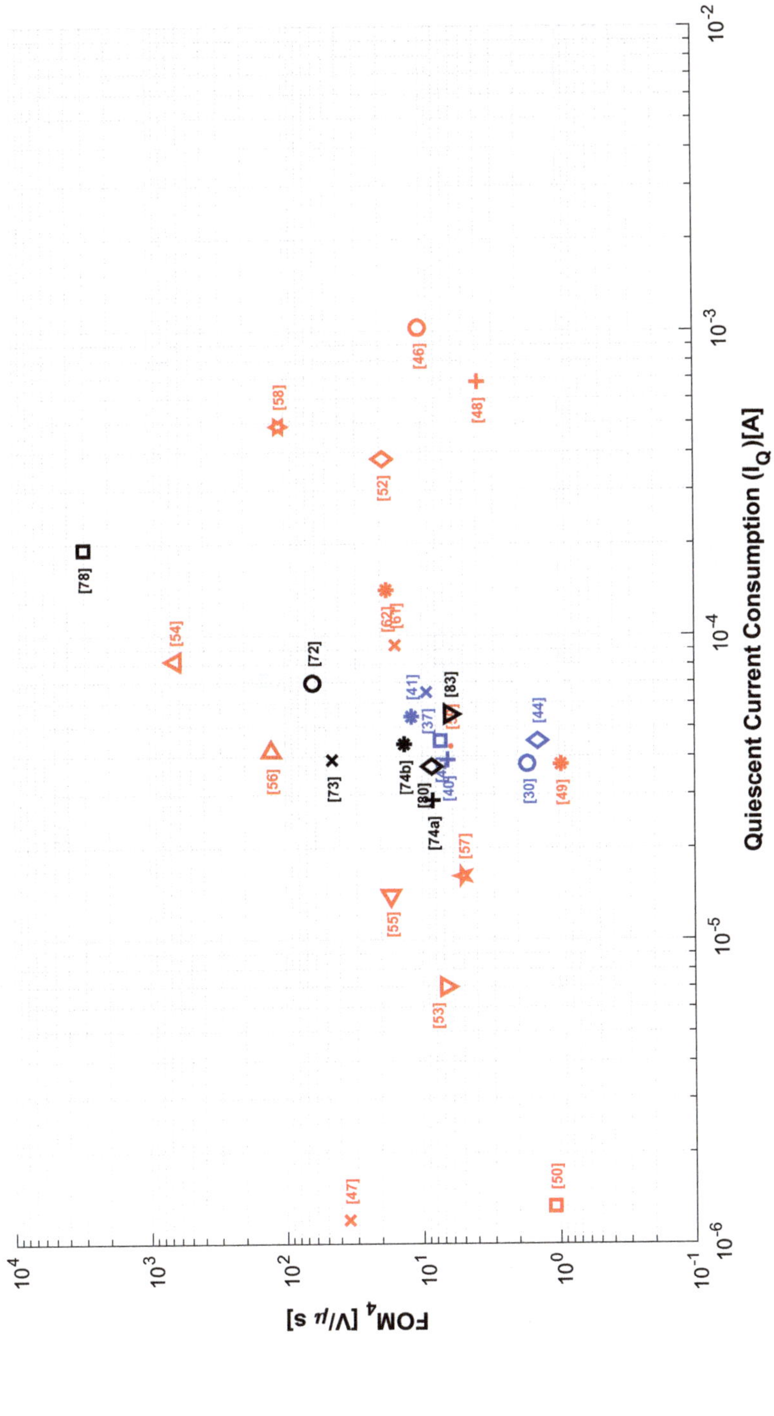

Fig. 2.27 FOM_4 versus quiescent current consumption

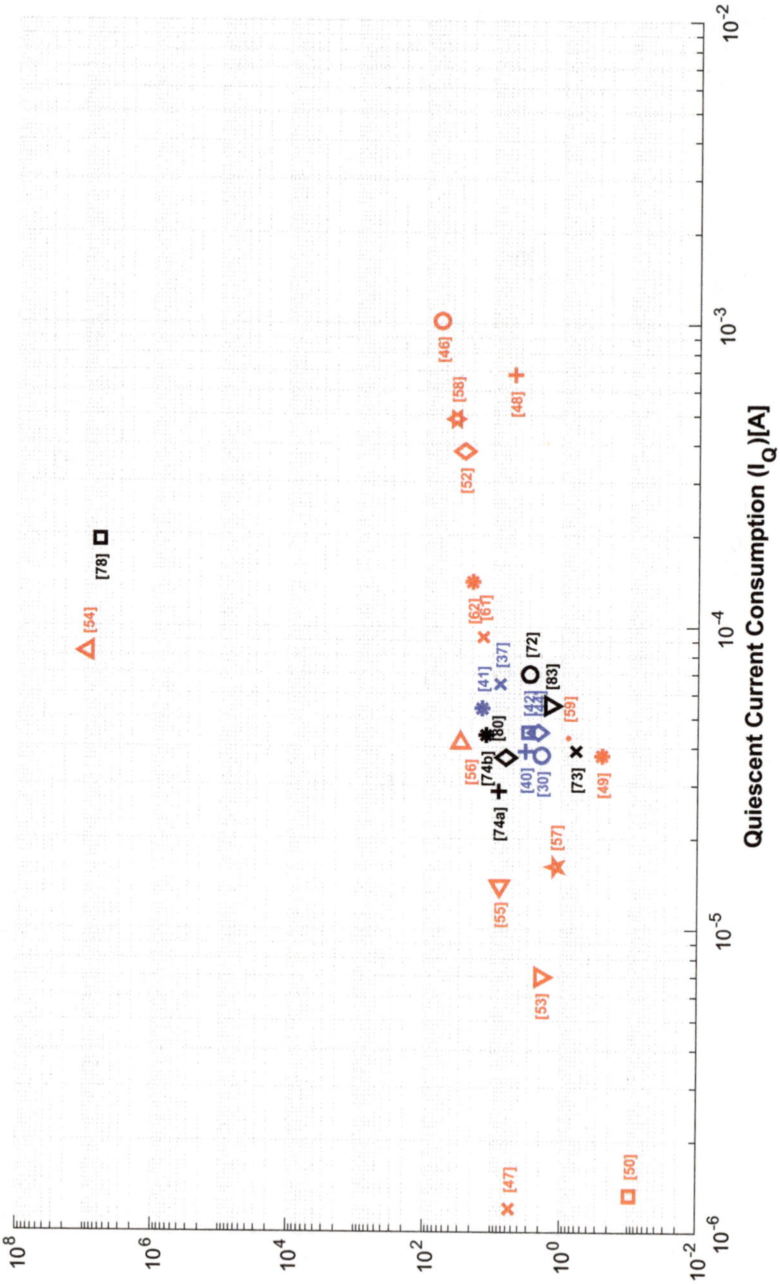

Fig. 2.28 FOM$_5$ versus quiescent current consumption

2.4 Conclusions of This Chapter

This chapter presents the classical topology of an IC-LDO and discusses the main problems affecting this structure.

Regarding stability, forcing the inner pole to be dominant, makes the non-dominant pole to be located at the output, and then, to be highly dependent on the load. Thus, when the load changes, the position of the non-dominant pole changes, which seriously affects the stability of the regulator. Therefore, it is necessary to study in detail the position of poles and zeros in different load ranges to ensure stability. Apart from the classic techniques to split poles and zeros, new solutions are discussed in this chapter. Most of these solutions are based on a control of the damping factor of the non-dominant poles.

The transient response is another important aspect in a regulator, since the presence of spikes at the output voltage can lead to a malfunction of the circuits connected to its output. The transient response is of special interest to internally compensated regulators, due to the absence of a large output capacitor that suppresses these spikes. A review of the techniques proposed to improve the transient response without penalizing the quiescent power consumption has been made. Special attention has been paid to those that involve the use of adaptive biasing, and the inclusion of output stages in the error amplifier to improve the slew-rate at the gate of the pass transistor.

The third main problem that affects LDO regulators is the effect that small variations of the input voltage has on the regulated output, which is characterized by the PSRR. The PSRR is of special importance to IC-LDO regulators used in SoCs, where analogue and digital circuits coexist. In the small-signal analysis, the behaviour of the PSRR in medium and high frequencies, where the main control loop is no longer effective, is particularly significant.

Finally, a comparison of the solutions proposed in the literature has been made, introducing some of the most relevant Figures of Merit.

Chapter 3
Adaptive Continuous Resistor for Miller Compensation in IC-LDO Regulators

Abstract IC-LDO regulators are circuits that play a key role in modern power management and SoC design. They are required to occupy small area, operate with low power consumption and low supply voltage, and perform fast transient response and good load and line regulations. In addition, they must remain stable under extreme variations of the input voltage, load current and load capacitance. In this chapter, an adaptive and continuous compensation technique is proposed that tunes the value of the zero-nulling resistor of a classical Miller-based compensation to keep the nulling zero close to the Unit Gain Frequency (UGF), according to the working conditions. This technique has been applied to an LDO regulator based on a classical topology. Results of an implementation of such a regulator fabricated in a standard 65 nm CMOS technology are shown at the end of the chapter.

3.1 Introduction

Stability is a key concern in electronic design, as it determines the performances of linear feedback systems [93]. In modern electronics, stability becomes technical challenge as, in addition to the high performance and efficiency that the market demands, the designer faces constraints coming from the downscaling of CMOS technologies [94].

The reduction of the supply voltage and the decrease of the intrinsic gain of MOS transistors [95, 96] owing to CMOS downscaling, highlight the relevance of low-power and low-voltage multistage circuits, whose close-loop response is degraded by the presence of multiple poles. At the same time, and as a consequence of local process variability and mismatch in modern CMOS technologies, integrated devices, such as resistors and capacitors, present large temperature variation and fabrication tolerances [97, 98]. In this scenario, classical compensation techniques are no longer valid, and designers need to devise new ones [99].

In particular, the IC-LDO regulator is attracting a lot of attention for new compensation methods, as it was shown in Chap. 2 (Sect. 2.2.1), where several compensation techniques were reviewed [30, 33, 37, 40–42, 44]. Most of them use active blocks in the signal path that increase the complexity and power consumption of the regulator,

© Springer International Publishing AG, part of Springer Nature 2018

J. M. Hinojo et al., *Internally Compensated LDO Regulators for Modern System-on-Chip Design*, https://doi.org/10.1007/978-3-319-75411-6_3

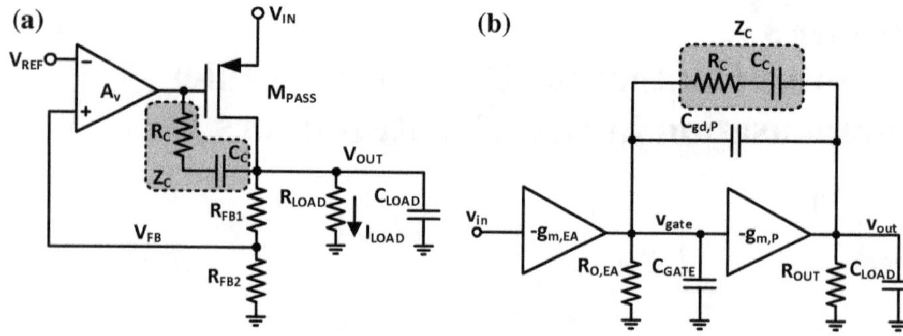

Fig. 3.1 **a** Classical topology of an IC-LDO regulator including the compensation network. **b** Open-loop small-signal model of a two-stage LDO regulator with Miller compensation

resulting in complex design equations, and large area or power consumption. In this chapter, one of the simplest compensation methods, the Miller one, is revisited, and a new technique is proposed to make it effective in IC-LDO regulators.

3.2 Analysis of the Miller Compensation Technique for IC-LDO Regulators

One of the easiest ways to increase the regulator stability is the so-called dominant pole compensation [93, 100]. The rationale behind this technique is to split the poles of the open-loop transfer function by placing a large capacitor at the selected node, which moves the associated pole towards lower frequencies, at the cost of a bandwidth reduction. In its simplest form, this technique is not suitable for SoCs, since such a large compensation capacitor cannot be integrated into a reasonable area.

However, if this idea is combined with the Miller effect, its efficient integration is possible [101]. In order to illustrate this technique, Fig. 3.1a shows an IC-LDO regulator with a classical topology, and Fig. 3.1b depicts its open-loop small-signal model, which is represented as a two-stage system. Note that the error amplifier is modelled by a single-pole opamp, while a medium-frequency small-signal model is used to represent M_{PASS}. This model is valid assuming the rest of poles from the error amplifier to be located at a high frequency.

In Fig. 3.1b, $C_{GATE} = C_{O,EA} + C_{gs,P}$ is the total capacitance between the gate of the pass transistor and the (small signal) ground. Capacitor $C_{gs,P}$ models the gate-to-source capacitance of M_{PASS} and $C_{gd,P}$ its gate-to-drain capacitance. $C_{O,EA}$ and $R_{O,EA}$ represent the output capacitance and resistance of the error amplifier, respectively. As usual, C_{LOAD} is the load capacitance at the output, and R_{OUT} is the equivalent resistance at the regulator output (Eq. 3.1).

$$R_{OUT} = r_{o,P} \| R_{LOAD} \| \left(R_{FB1} + R_{FB2}\right) \tag{3.1}$$

Initially, the compensation impedance Z_C will be assumed to be purely capacitive (i.e. $R_C = 0\,\Omega$). Then, the open-loop transfer function is given by

$$H_{MC}(s) = A_{OL}\frac{1 - s\frac{C_C + C_{gd,P}}{g_{m,P}}}{1 + sd_1 + s^2 d_2} \tag{3.2}$$

where

$$A_{OL} = g_{m_{EA}} g_{m,P} R_{O,EA} R_{OUT} \tag{3.3}$$

$$d_1 = R_{O,EA}\left[C_{GATE} + \left(1 + g_{m,P}R_{OUT}\right)\left(C_C + C_{gd,P}\right)\right] + + R_{OUT}\left(C_{LOAD} + C_C + C_{gd,P}\right) \tag{3.4}$$

$$d_2 = R_{O,EA}R_{OUT}\left[C_{GATE}\left(C_C + C_{gd,P} + C_{LOAD}\right) + C_{LOAD}\left(C_C + C_{gd,P}\right)\right] \tag{3.5}$$

In order to simplify these equations, it will be assumed that $C_{gd,P} \ll C_C \ll C_{LOAD}$ and $gm_{EA}R_{O,EA} \gg 1$, leading to Eq. 3.6.

$$H(s) \approx \frac{A_{OL}\left(1 - s\frac{C_C + C_{gd,P}}{g_{m,P}}\right)}{1 + s\left(R_{OUT}C_{LOAD} + R_{O,EA}\left[C_{GATE} + C_C\left(1 + g_{m,P}R_{OUT}\right)\right]\right) + s^2 R_{O,EA}R_{OUT}C_{LOAD}\left(C_C + C_{GATE}\right)} \tag{3.6}$$

If the Miller compensation is properly designed, there will be a dominant pole $\omega_{p1} \ll \omega p2$, and one zero $\omega z1$:

$$H_{MC}(s) \approx A_{OL}\frac{\left(1 - \frac{s}{\omega_{z1}}\right)}{\left(1 + \frac{s}{\omega_{p1}}\right)\left(1 + \frac{s}{\omega_{p2}}\right)} \approx \frac{1 - \frac{s}{\omega_{z1}}}{1 + s\frac{1}{\omega_{p1}} + s^2\frac{1}{\omega_{p1}\omega_{p2}}} \tag{3.7}$$

$$\omega_{p1} \approx \frac{1}{R_{O,EA}\left(C_{GATE} + g_{m,P}R_{OUT}C_C\right)} \tag{3.8}$$

$$\omega_{p2} \approx \frac{1}{R_{OUT}C_{LOAD}\frac{C_C + C_{GATE}}{C_{GATE} + g_{m,P}R_{OUT}C_C}} \tag{3.9}$$

$$\omega_{z1} \approx \frac{g_{m,P}}{C_C + C_{gd,P}} \tag{3.10}$$

Identifying terms with Eq. 3.6, it can be concluded that the system has two real poles (Eqs. 3.8 and 3.9) and one Right Half-Plane (RHP) zero (Eq. 3.10). The latter zero depends on the ratio between $g_{m,P}$ and the total Miller capacitance $C_C + C_{gd,P}$. Unfortunately, as both, $g_{m,P}$ and $C_{gd,P}$ are strongly dependent on I_{LOAD}, the RHP zero location changes several decades when the I_{LOAD} is modified, moving towards

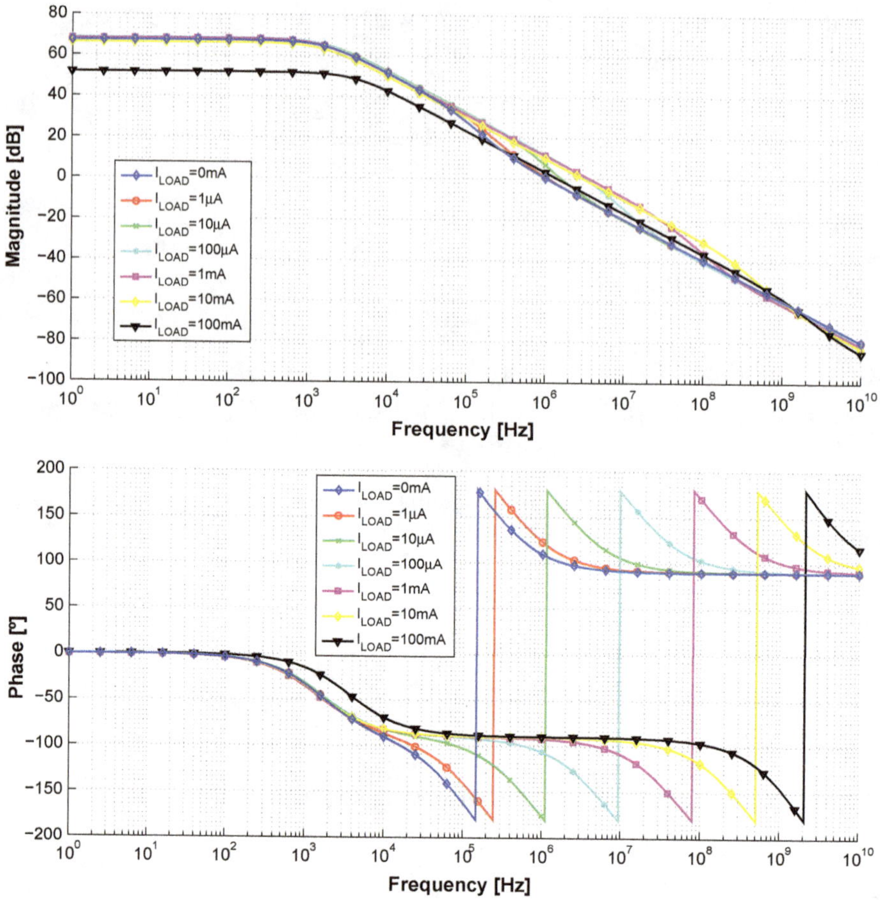

Fig. 3.2 Open-loop transfer function for an IC-LDO regulator stabilized using Miller compensation

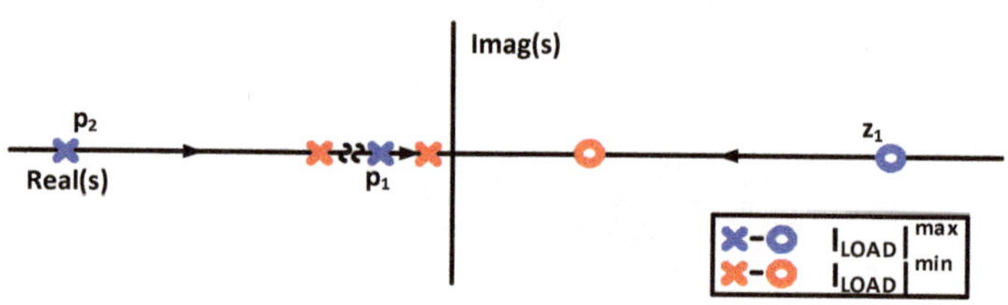

Fig. 3.3 Pole-zero diagram for different values of I_{LOAD} in a typical case

the Unity Gain Frequency (UGF) and, hence, degrading the phase margin, as it is shown in Figs. 3.2 and 3.3.

In order to neutralize the negative effects of this RHP zero, different solutions have been proposed in the literature, such as the well-known Zero-Nulling Resistor

(ZNR) technique [93], the cascode compensation [102] or the insertion of a current buffer [34]. The first option cancels the RHP zero effect by means of a nulling resistor (R_C) in series with the compensation capacitor C_C. Unfortunately, as it is strongly dependent on I_{LOAD}, the value of R_C that stabilizes the LDO regulator for zero-load current is not adequate for middle-range load currents, as it will be shown in Sect. 3.3. As a consequence, the LDO regulator is not stable for the whole range of I_{LOAD}. Some authors use a current buffer to eliminate this RHP zero [35, 37]. Nevertheless, this solution increases the power consumption and the complexity of the design, and it does not guarantee the stability for $I_{LOAD} = 0\,A$.

A more realistic model of the regulator would show the presence of two complex non-dominant poles. Therefore, some authors propose to control the Damping Factor (DF) of these non-dominant complex poles to achieve an effective pole-splitting compensation [30, 33, 103]. Nevertheless, these LDO regulators are potentially unstable for low I_{LOAD} and require a high quiescent consumption. Another option was proposed in [44] where an RNMC technique based on current buffers introduces two Left Half-Plane (LHP) zeros that cancel one of the non-dominant poles of the system. This approach requires a large capacitor to stabilize the regulator and it is not suitable for I_{LOAD} values close to zero.

When the error amplifier is a multistage amplifier the Nested Miller Compensation (NMC) is a choice [48, 62], at the cost of an even more reduced bandwidth, as each stage is loaded not only by its output parasitic capacitance but also by the Miller compensation capacitors. Note that additional circuitry and power consumption are required to maintain the Slew-Rate (SR), and in turn, the transient response of the LDO regulator. In addition, the NMC limits the architecture of the error amplifier, as its last stage must be a non-inverting one in order to ensure a negative feedback [100]. Finally, as a consequence of the large variation in the operating conditions, a thorough analysis of the frequency response must be done to determine the proper value of the NMC capacitors [104–106].

Summarizing, the stability of IC-LDO regulator is a challenge that implies complex compensation techniques that result in large area or power consumption. In this paper, a new solution is proposed that uses a simple and well-known compensation method, such as the classical Miller one with ZNR, but adapting the value of the nulling resistor to the variations of I_{LOAD}.

3.3 Stability Analysis of Miller-Compensated IC-LDO Regulators

A deep analysis of the zero-nulling resistor in a Miller-based compensation of IC-LDO regulators is done in this section. The analysis will target the stability for a wide range of load currents, including zero-load current. As it was discussed in Chap. 2, it is focused on I_{LOAD} as V_{IN} is not relevant for the stability.

When the compensation resistor, R_C, is included in the analysis of the small-signal model of Fig. 3.1b, the resulting open-loop transfer function is

$$H_{ZNR}(s) = A_{OL}\frac{1 + s\left(R_C C_C - \frac{C_C + C_{gd,P}}{g_{m,P}}\right) - s^2\frac{R_C C_C C_{gd,P}}{g_{m,P}}}{1 + sd_1 + s^2 d_2 + s^3 d_3} \tag{3.11}$$

where

$$A_{OL} = g_{m,EA} g_{m,P} R_{O,EA} R_{OUT} \tag{3.12}$$

$$R_1 = \left(R_{O,EA} + R_{OUT}\right)\left(1 + g_{m,P} R_{O,EA} \| R_{OUT}\right) \tag{3.13}$$

$$d_1 = \left(R_C + R_1\right) C_C + R_1 C_{gd,P} + R_{OUT} C_{LOAD} + R_{O,EA} C_{GATE} \tag{3.14}$$

$$\begin{aligned} d_2 = R_{O,EA} R_{OUT} \Bigg(& C_{LOAD}\left[C_{gd,P} + C_{GATE} + \left(1 + \frac{R_C}{R_{O,EA}}\right) C_C\right] + \\ & + C_{GATE}\left[C_{gd,P} + C_{GATE} + \left(1 + \frac{R_C}{R_{OUT}}\right) C_C\right]\Bigg) + R_C R_1 C_C C_{gd,P} \end{aligned} \tag{3.15}$$

$$d_3 = R_C R_{O,EA} R_{OUT} C_C \left[C_{LOAD}\left(C_{gd,P} + C_{GATE}\right) + C_{gd,P} C_{GATE}\right] \tag{3.16}$$

A quantitative analysis of the transfer function for different values of I_{LOAD} is presented in the following subsections. This will provide relevant information on how R_C modifies the frequency response (Eqs. 3.11–3.16) and, consequently, affects the stability of the system.

In what follows, as an example, this analysis is made for the IC-LDO regulator designed in Sect. 3.5 using a standard 65-nm CMOS technology. Although the analysis is made for a given example, its conclusions can be extrapolated to a more general case.

For the regulator in Sect. 3.5, Fig. 3.4 depicts the Bode diagram for different values of I_{LOAD}, and Fig. 3.5 presents the pole-zero diagram showing the movement of poles and zeros when I_{LOAD} changes. These figures have been obtained for a fixed value of the zero-nulling resistor.

3.3.1 *Zero or Low* I_{LOAD}

For I_{LOAD} from 0 up to 50–100 μA, M_{PASS} is working in sub-threshold region. Therefore, it can be assumed that $g_{m,P}$ is in the order of tens of $\mu\Omega^{-1}$, R_{OUT} is around tens or hundreds of kΩ, and consequently $g_{m,P}R_{OUT} \gg 1$, Fig. 3.6. In addition, $C_{O,EA} \ll C_{gd,P}, C_{LOAD}$. Thus, the transfer function can be approached as follows:

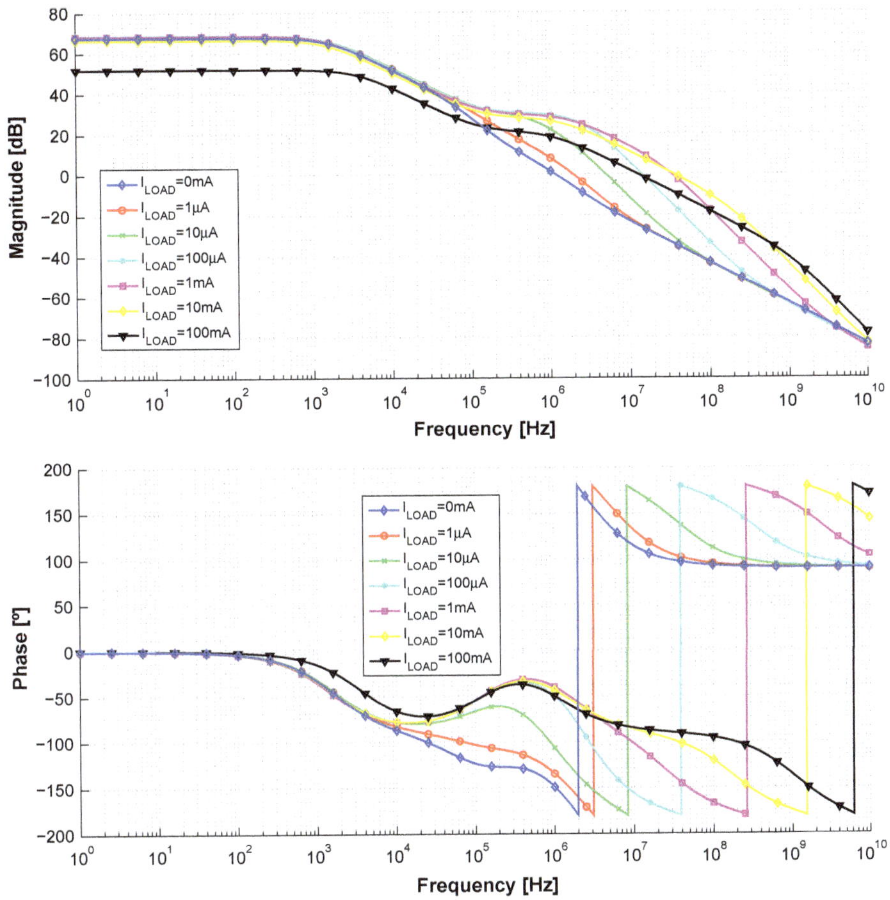

Fig. 3.4 Bode diagram for an LDO regulator with classical Miller compensation and ZNR

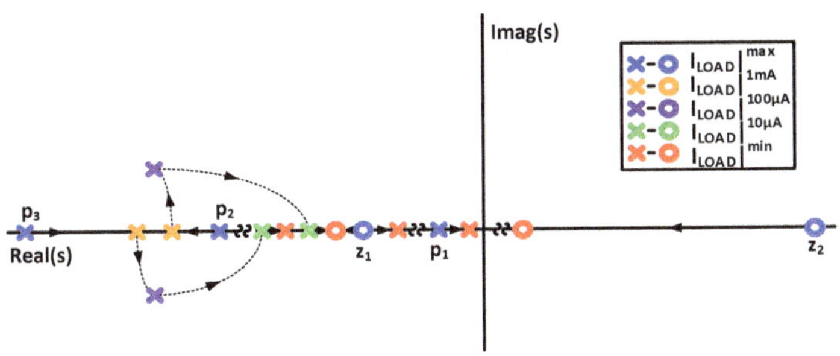

Fig. 3.5 Pole-zero diagram for different values of I_{LOAD} for the regulator in Sect. 3.5, $C_{LOAD} = 100\,pF$

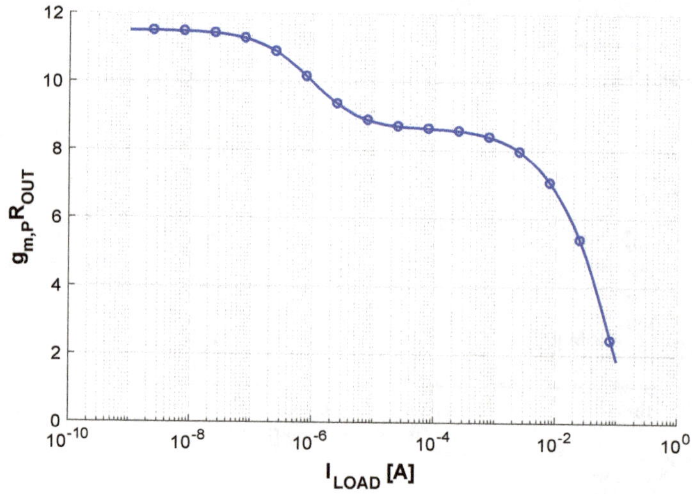

Fig. 3.6 $g_{m,P}$ R_{OUT} for different I_{LOAD} values

$$H_{ZNR}(s) = A_{OL}\frac{1+\left(R_C C_C - \frac{C_C+C_{gd,P}}{g_{m,P}}\right)s - \frac{R_C C_C C_{gd,P}}{g_{m,P}}s^2}{\left(1+\frac{s}{\omega_{p1}}\right)\left(1+d_1 s + d_2 s^2\right)} \tag{3.17}$$

where

$$A_{OL} = g_{m,EA} g_{m,P} R_{O,EA} R_{OUT} \tag{3.18}$$

$$\omega_{p1} \approx \frac{1}{R_{O,EA}\left[C_{O,EA} + g_{m,P} R_{OUT}\left(C_C + C_{gd,P}\right)\right]} \tag{3.19}$$

$$c_1 = \frac{C_{LOAD}}{g_{m,P}} + \frac{R_C C_C C_{gd,P}}{C_C + C_{gd,P}} \tag{3.20}$$

$$c_2 = \frac{R_C C_C C_{gd,P} C_{LOAD}}{g_{m,P}\left(C_C + C_{gd,P}\right)} \tag{3.21}$$

In order to stabilize the IC-LDO regulator, R_C and C_C are chosen, using Eqs. 3.22 and 3.23, to obtain a PM large enough for $I_{LOAD} = 0$ A, which is the worst case in the selected range. Indeed, R_C is calculated to locate an LHP zero close to the UGF.

$$PM = tan^{-1}\left(\frac{1 - c_2\omega_{UGF} + \left(R_C C_C - \frac{C_C+C_{gd,P}}{g_{m,P}}\right)c_1\omega_{UGF}^2}{\omega_{UGF}\left(1 - c_2\omega_{UGF}\right)\left[c_1 - \left(R_C C_C - \frac{C_C+C_{gd,P}}{g_{m,P}}\right)\right]}\right) \tag{3.22}$$

$$R_C > \frac{C_C + C_{gd,P}}{g_{m,P} C_C} \tag{3.23}$$

Curves in Fig. 3.4 corresponding to I_{LOAD} from 0 A up to 10 μ A show the effect of the LHP zero introduced by the ZNR, and the location of the non-dominant poles for small values of I_{LOAD}. Furthermore, Fig. 3.5 visualizes the pole-zero movement when I_{LOAD} changes.

3.3.2 Medium I_{LOAD}

When I_{LOAD} reaches 100 μA, and up to 10 mA, M_{PASS} is working in moderate to strong inversion, in saturation region. Like in the previous scenario, the assumption $g_{m,P}R_{OUT} \gg 1$ is acceptable, as the $g_{m,P}$ value is around hundreds of $\mu\Omega^{-1}$ and the output resistance of M_{PASS} is around several kΩ (Fig. 3.6). The transfer function in Eq. 3.17 and the following equations (Eqs. 3.18–3.23) are still valid. Under these conditions, the Q-factor of the second-order polynomial increases, producing non-dominant complex poles that shift the phase 180° in one decade. As the location of the LHP zero is fixed, the UGF moves towards a frequency higher than those of the non-dominant complex poles. This implies a significant reduction of the PM. This situation is depicted in Fig. 3.6 for I_{LOAD} between 100 μA and 10 mA.

3.3.3 Large I_{LOAD}

For load currents larger than tens of milliamps, the M_{PASS} transistor is working in the triode region and therefore, the transconductance $g_{m,P}$ is in the order of a few (Ω^{-1}), whereas the output resistance R_{OUT} is around a few ohms. Consequently, $g_{m,P}R_{OUT} \approx 1$ (see Fig. 3.6) and Eq. 3.17 is no longer valid. In this case, the system presents two non-dominant real poles and the LHP zero introduced by the compensation. The LHP zero is located below the first non-dominant pole, cancelling its effect. The second non-dominant pole is located beyond the UGF. Consequently, the regulator is stable. This situation is depicted in Fig. 3.4 for I_{LOAD} between 10 and 100 mA.

Figure 3.7 represents the PM for different values of I_{LOAD} in the cases of an uncompensated IC-LDO regulator, and a compensated Miller version with, and without, a ZNR. It can be observed that there is range of values of I_{LOAD} where the PM is not large enough.

From the previous analysis, it can be inferred that only a change in the value of the nulling resistor when I_{LOAD} changes would guarantee the stability of the IC-LDO regulator in the whole operation region. This is the rationale behind the compensation method proposed in this chapter.

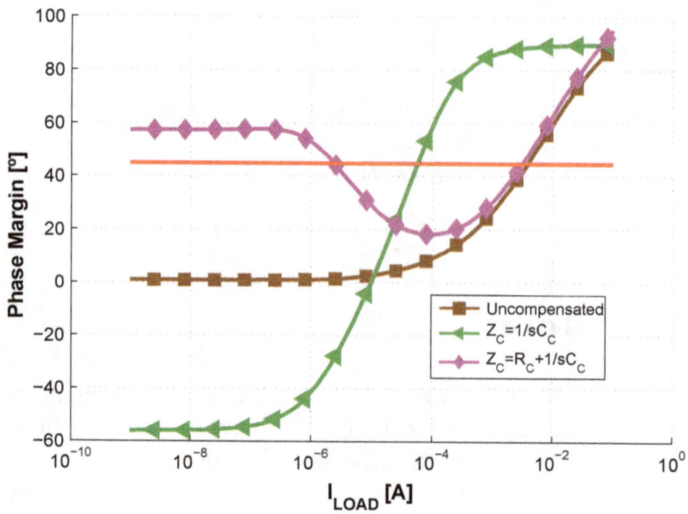

Fig. 3.7 PM variation vs. I_{LOAD} for an uncompensated IC-LDO regulator and a Miller-compensated one with, and without, a ZNR

3.4 Miller Compensation with Continuous Adaptive Zero-Nulling Resistor

The stability of uncompensated IC-LDO regulators and the suitability of Miller compensation have been studied in Sects. 2.2.1 and 3.3, respectively. A similar analysis could be done for some other Miller-compensated analogue cells intended to manage fast and significant load variations. Figure 3.8 shows a general control scheme, where IN_X signals are the external conditions that play a role in determining the appropriate value of the nulling resistor. The control block selects this value both, in a continuous or discrete way. When possible, a smooth variation of R_C is preferred to avoid an undesirable transient response in the switching instants, which could cause instability.

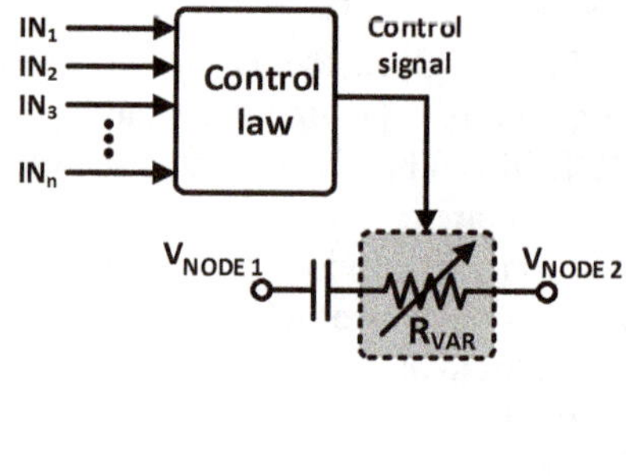

Fig. 3.8 Proposed control scheme for the modification of the nulling resistor in a classical Miller compensation with ZNR

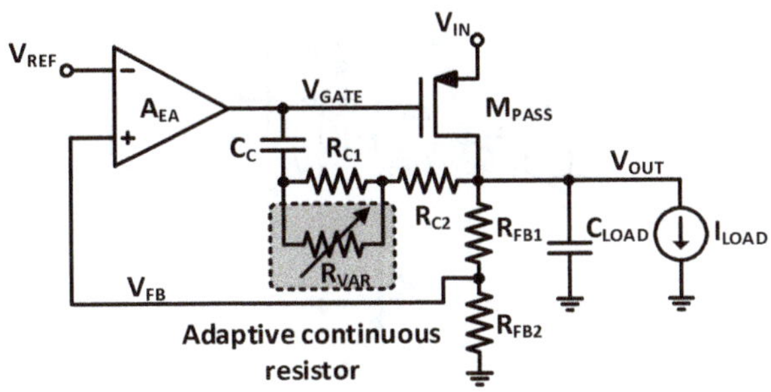

Fig. 3.9 Proposed compensation technique for a classical IC-LDO regulator

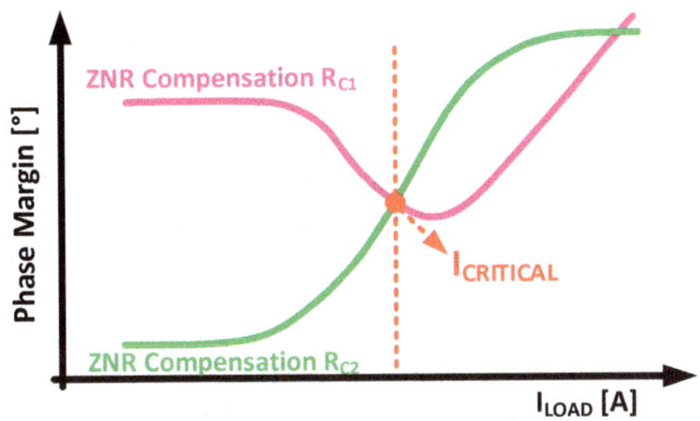

Fig. 3.10 Graphical determination of the critical I_{LOAD}

According to the results obtained in Sect. 2.2.1, the only external condition to be considered in an IC-LDO regulator is I_{LOAD}. This simplifies the implementation of the control loop and makes this technique especially suitable for IC-LDO regulators.

Following the control scheme of Figs. 3.8 and 3.9 depicts the proposed compensation scheme for an IC-LDO regulator. In order to achieve a practical implementation, the nulling resistor R_C is split into three different resistors: R_{C1}, R_{C2} and R_{VAR}. Resistor R_{C1} is selected to guarantee the stability of the LDO regulator for $I_{LOAD} = 0$ A, whereas R_{C1} ensures that an LHP zero is moved beyond UGF to stabilize the LDO regulator at high I_{LOAD} values. Note that $R_{C2} \ll R_{C1}$. Finally, R_{VAR} is a variable resistance implemented by M_{VAR}, a transistor in the triode region.

This implementation of R_C relaxes the accuracy requirements of the control circuit, as the values in the extremes of the interval are well defined by R_{C1} and R_{C2}, respectively. Regarding R_{VAR}, its nominal value is selected to ensure that the value of the equivalent R_C stabilizes the IC-LDO regulator for a specific value of I_{LOAD}, named the $I_{CRITICAL}$ (Fig. 3.10), which is defined as the I_{LOAD} for which the PM of the IC-LDO regulator with $R_C = R_{C1}$ is equal to that obtained with $R_C = R_{C2}$.

Figure 3.11 shows the proposed control circuit, where V_{OUT} is the regulated output voltage and M_{PR} and I_{CR} are scaled versions of the pass transistor and $I_{CRITICAL}$, respectively. Additionally, A_1 and A_2 are two NMOS-input differential pairs with active current mirrors. Finally, a level shifter, implemented by M_{LS} and $I_{BIAS,LS}$, is

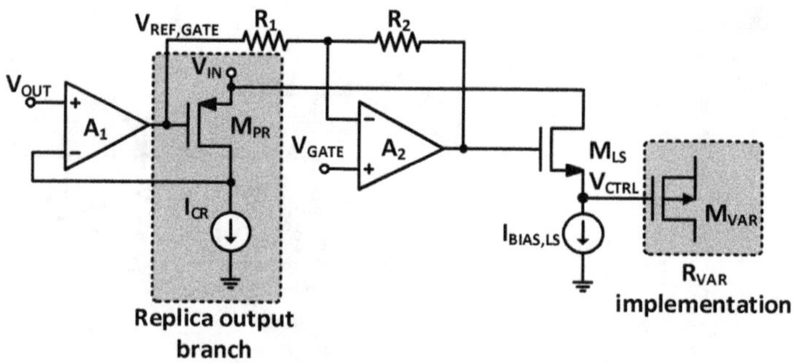

Fig. 3.11 Circuit proposed to control the adaptive continuous resistor value by means of the gate voltage of M_{VAR}

needed to adapt the output signal of the differential pair A_2 to the range of the gate voltages required by the transistor M_{VAR}.

Assuming the virtual ground principle in A_1, the drain voltage in M_{PR} is forced to be equal to V_{OUT} and, as the current of M_{PR} is set by I_{CR}, the output voltage of A_1 generates the reference voltage ($V_{REF,GATE}$) that corresponds to the LDO regulator conditions for the critical I_{LOAD}. This $V_{REF,GATE}$ is, then, used in A_2 to be compared to the V_{GATE} of the LDO core. A_2 output is adapted by a level shifter to provide the control voltage V_{CTRL}. This voltage is modified according to Eq. 3.24 in order to tune the output resistance of M_{VAR} to a proper value. R_2/R_1 is selected according to the desired accuracy and M_{VAR} is sized using Eq. 3.25 [107–109].

$$V_{CTRL} = \left(1 + \frac{R_2}{R_1}\right) V_{GATE} - \frac{R_2}{R_1} V_{REF,GATE} - V_{EFF,M_{LS}} \tag{3.24}$$

$$\left.\frac{W}{L}\right|_{M_{VAR}} = \frac{1}{2n\mu C_{OX} U_t^2 R_{VAR}} exp\left(-\frac{V_{CTRL} - V_{t0}}{nU_t}\right) \tag{3.25}$$

where U_t is the thermal voltage; and n, μ, C_{OX} and V_{t0} are the slope factor, mobility, gate capacitance per unit area and threshold voltage of M_{VAR}, respectively.

3.5 Design of an IC-LDO Regulator with Continuous Adaptive Zero-Nulling Resistor

This section describes an LDO regulator designed to verify the proposed technique. It has been implemented in a standard 65 nm CMOS technology and designed to provide up to 100 mA of I_{LOAD} with a minimum input voltage of 1.0 V. The output voltage is 0.8 V with a dropout voltage of 200 mV, and 100 pF maximum load capacitance.

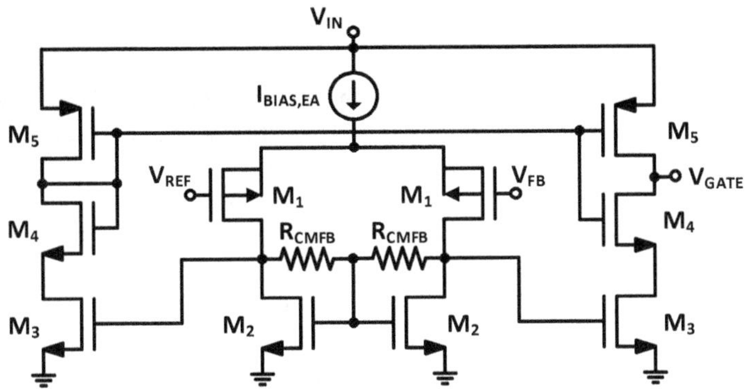

Fig. 3.12 Schematic of the error amplifier used in the proposed IC-LDO regulator

Table 3.1 Transistor dimensions and component values for the LDO regulator core

Transistor	Size [μm/μm]
M_{PASS}	2000.0/0.06
M_1	2.7/0.27
M_2	0.54/0.27
M_3	6.51/0.27
M_4	7.53/0.27
M_5	8.1/0.27
Component	Value
R_{CMFB}	200 [kΩ]
R_{FB1}	300 [kΩ]
R_{FB2}	300 [kΩ]
$I_{BIAS,EA}$	2.5 [μA]
$C_{LOAD,Max}$	100 [pF]
C_C	18.48 [pF]

The classical LDO topology with the proposed Miller compensation and continuous adaptive nulling resistor of Fig. 3.11 has been selected. The Error Amplifier is implemented by means of the single-ended two-stage amplifier shown in Fig. 3.12. This amplifier has been used before to implement the EA of an LDO regulator [83] showing a class-AB behaviour.

The design methodology assumes that the dominant pole is located at the gate of M_{PASS} and the non-dominant pole is at the output. Additionally, the error amplifier is designed to ensure that its internal poles are at a frequency high enough so as not to compromise the overall stability. Table 3.1 summarizes the transistors dimensions and device values for the LDO core.

3.5.1 Stability Analysis

In Sect. 3.4, a methodology for the design of the proposed compensation network was described, which is based on the determination of a critical I_{LOAD}. A stability analysis of the designed LDO regulator is performed in this section. Figure 3.13 depicts its small-signal model, where $g_{m,1}$ and $g_{m,2}$ are the transconductance of the first and the second stages of the error amplifier, respectively. The output resistance of the first stage is represented by $R_{1stg,EA}$ and $C_{1stg,EA}$ models the parasitic capacitance at the gate of transistor M_3 in Fig. 3.12. It is important to highlight that the replica circuit that controls the gate voltage of M_{VAR} is not in the signal path and, consequently, the whole compensation scheme can be modelled as a passive network composed of three single resistors, R_{VAR}, R_{C1} and R_{C2}, and the capacitor, C_C.

The frequency compensation of the open-loop response is obtained by means of the Miller capacitor, C_C, and the proposed adaptive continuous resistor that makes possible the movement of the LHP zero. Assuming $g_{m,1}R_{1stg,EA}$, $g_{m,2}R_{O,EA} \gg 1$ and $C_{LOAD} \gg C_{GATE} \gg C_{1stg,EA}$, the frequency response of the designed IC-LDO regulator can be approached by Eq. 3.26, where the DC gain, A_{OL}, and the dominant pole, Ω_{p1}, are given by Eqs. 3.27 and 3.28, respectively.

$$H(s) = A_{OL}\frac{1 + sa_1 + s^2a_2}{\left(1 + \frac{s}{\omega_{p1}}\right)\left(1 + sb_1 + s^2b_2\right)} \tag{3.26}$$

$$A_{OL} = g_{m,1}R_{1stg,EA}g_{m,2}R_{O,EA}g_{m,P}R_{OUT} \tag{3.27}$$

$$\omega_{p1} \cong \frac{1}{R_{O,EA}\left[C_{O,EA} + g_{m,P}R_{OUT}\left(C_C + C_{gd,P}\right)\right]} \tag{3.28}$$

The remaining coefficients, related to zeros and non-dominant complex poles, are given by Eqs. 3.29–3.32.

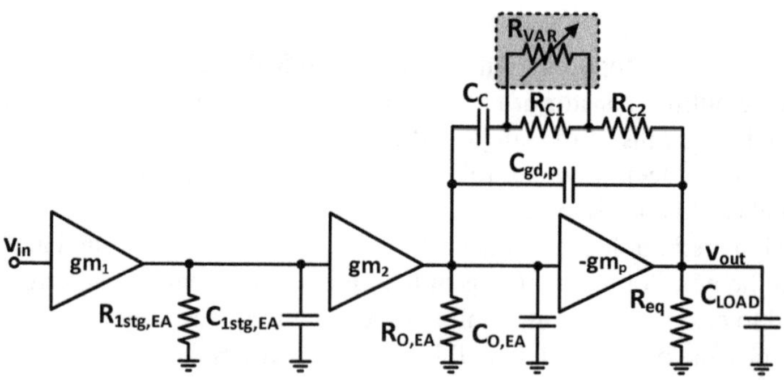

Fig. 3.13 Small-signal model of the proposed IC-LDO regulator

$$a_1 = \frac{C_C + C_{gd,P} - g_{m,P} R_C C_C}{g_{m,P}} \tag{3.29}$$

$$a_2 = \frac{R_C C_C C_{gd,P}}{g_{m,P}} \tag{3.30}$$

$$b_1 = \frac{R_{OUT}\left[C_{LOAD}\left(C_{GATE} + C_{gd,P} + C_C\right) + C_{GATE}\left(C_{gd,P} + C_C\right)\right]}{C_{O,EA} + g_{m,P} R_{OUT}\left(C_{gd,P} + C_C\right)} \tag{3.31}$$

$$b_2 = \frac{R_{OUT} R_C C_C\left[C_{LOAD}\left(C_{GATE} + C_{gd,P}\right) + C_{GATE} C_{gd,P}\right]}{C_{O,EA} + g_{m,P} R_{OUT}\left(C_{gd,P} + C_C\right)} \tag{3.32}$$

According to Eqs. 3.26–3.32, the phase margin (PM) can be calculated as

$$PM \approx tan^{-1}\left[\frac{1 - (a_2 + b_2 - a_1 b_1)\,\omega_{UGF}^2 + a_2 b_2 \omega_{UGF}^4}{(a_1 - b_1)\omega_{UGF} - (a_1 b_2 - a_2 b_1)\omega_{UGF}^3}\right] \tag{3.33}$$

Applying the design methodology described in the previous section, R_{C1} is selected to compensate the LDO regulator at $I_{LOAD} = 0\,A$ and R_{C2} at $I_{LOAD} = 100\,mA$, resulting in $70\,k\Omega$ and 200Ω, respectively. d Simulating the PM for the LDO regulator using the calculated R_{C1} and R_{C2}, $I_{CRITICAL}$ can be estimated to be $14\,\mu A$.

Regarding the control circuit, as M_{PASS} is implemented by 200 unitary transistors, the scale ratio is set to $K = 200$ in order to ensure that M_{PR} accurately replicates the bias point of M_{PASS}. Thus, the size of M_{PR} will be $10/0.06\,\mu m$ and $I_{CR} = 70\,nA$. Additionally, the desirable value of R_C at $I_{CRITICAL}$ can be estimated using Eq. 3.33. Finally, from $R_C = R_{VAR} \| (R_{C1} + R_{C2})$ and Eq. 3.25, the nominal value of R_{VAR} and the size of M_{VAR} are selected. Table 3.2 summarizes the transistor dimensions and device values for the control circuit, as well as for the adaptive nulling resistor. Figure 3.14a depicts the value of $R_{VAR} = r_{ds}|^{M_{VAR}}$ versus I_{LOAD}, while Fig. 3.14b represents the equivalent nulling resistor, R_C.

Table 3.2 Transistor dimensions and component values for the control circuit

Transistor	Size [μm/μm]
M_{PR}	10.0/0.06
M_{VAR}	13/0.06
Component	Value
R_{C1}	70 [kΩ]
R_{C2}	200 [Ω]
R_1	100 [kΩ]
R_2	1 [MΩ]
I_{CR}	70 [nA]

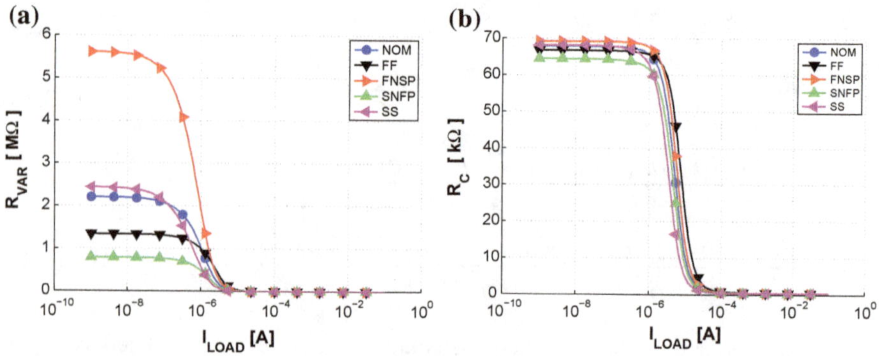

Fig. 3.14 Variation of **a** R_{VAR}, and **b** R_C, versus I_{LOAD}

Table 3.3 Simulated phase margin and gain versus I_{LOAD}

I_{LOAD}	PM [°]	Gain $[dB]$	I_{LOAD}	PM [°]	Gain $[dB]$
0 [nA]	61.39	49.33	10 [μA]	57.59	55.45
1 [nA]	61.41	49.33	100 [μA]	70.78	57.29
10 [nA]	61.63	49.37	1 [mA]	89.87	58.27
100 [nA]	63.28	49.71	10 [mA]	92.16	58.38
1 [μA]	66.75	51.85	100 [mA]	92.85	27.94

3.5.2 *Simulations*

This circuit has been designed in a standard 65 nm CMOS technology. Figure 3.15 shows post-layout simulations of the open-loop response for different I_{LOAD} conditions (from 0 to 100 mA). Simulations have been carried out assuming the worst case regarding stability, i.e. $V_{IN} = 1.0\,V$ and $C_{LOAD} = 100\,pF$. Post-layout simulated gain and phase margin are summarized in Table 3.3, respectively. Additionally, Fig. 3.16 represents the change in the location of poles and zeros when the load current changes.

Moreover, in order to clarify these results, Fig. 3.17 depicts the variation of the PM versus I_{LOAD} showing that, from 1 μA to 1 mA, the effect of the variable resistance increases the PM, avoiding the critical region that appears when the value of the resistor is fixed (Fig. 3.7).

In order to demonstrate the robustness of the proposed technique, process and mismatch Monte Carlo simulations for different load currents have been done. Figures 3.18–3.20 show the results obtained for the phase margin, while Table 3.4 summarizes the statistical information, i.e. mean value and standard deviation, for each load condition.

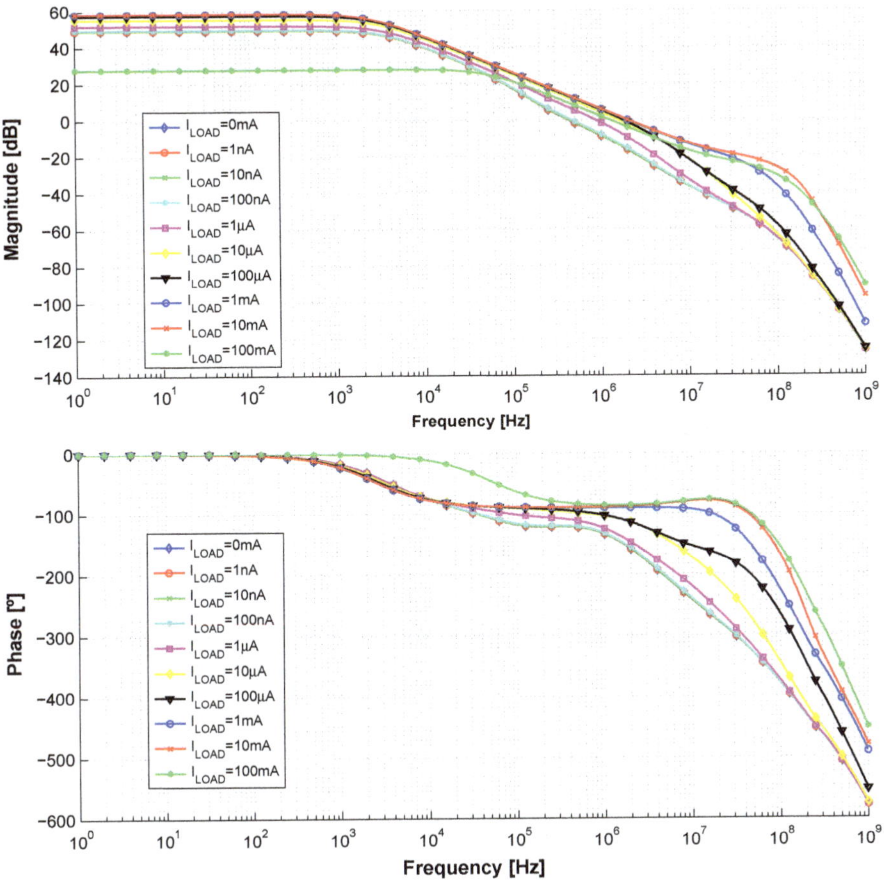

Fig. 3.15 Simulated open-loop response of the proposed IC-LDO regulator for different values of I_{LOAD}

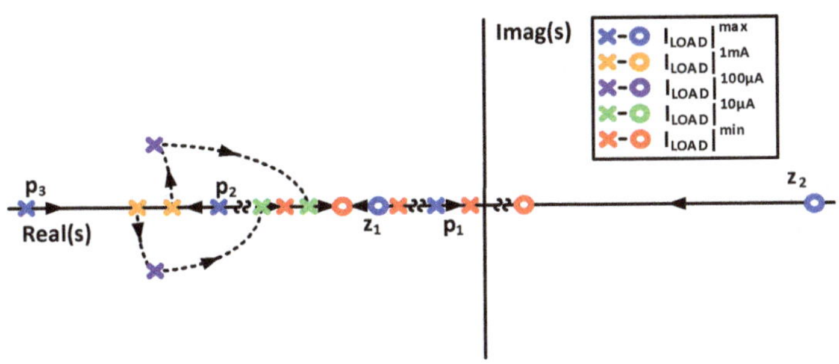

Fig. 3.16 Simulated pole-zero diagram for the proposed IC-LDO regulator

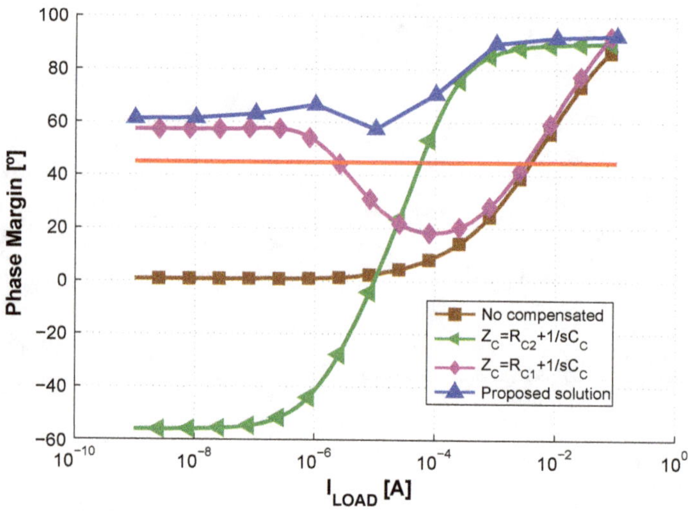

Fig. 3.17 Comparison of the proposed frequency compensation technique, the uncompensated version and the classical Miller compensation with, and without, ZNR

Table 3.4 Simulated post-layout gain and phase margin for different load conditions

I_{LOAD}	Mean PM [°]	σ [°]
0 [nA]	69.82	4.07
100 [nA]	61.63	4.19
1 [μA]	66.50	4.36
10 [μA]	54.66	16.20
100 [μA]	75.50	6.78
1 [mA]	90.47	1.41
10 [mA]	92.07	0.55
100 [mA]	93.05	1.29

3.5.3 *Experimental Results*

Figure 3.21 shows the layout of the IC-LDO regulator superimposed on a multi-project chip microphotograph, where region *A* represents the pass transistor, region *B* corresponds to the error amplifier, region *C* shows the implementation of the control circuit, and region *D* is a biasing circuit that generates the required bias voltages and currents. C_C is the Miller capacitance; whose value is 18.48 pF. The total area is $289 \times 151\,\mu$m, where the control circuit occupies an area of $46 \times 30.2\,\mu$m and its current consumption is 1.8 μA.

Once the fabricated samples were received, they were measured with the test setup presented in Appendix B. Figures 3.22 and 3.23 depict the measured load

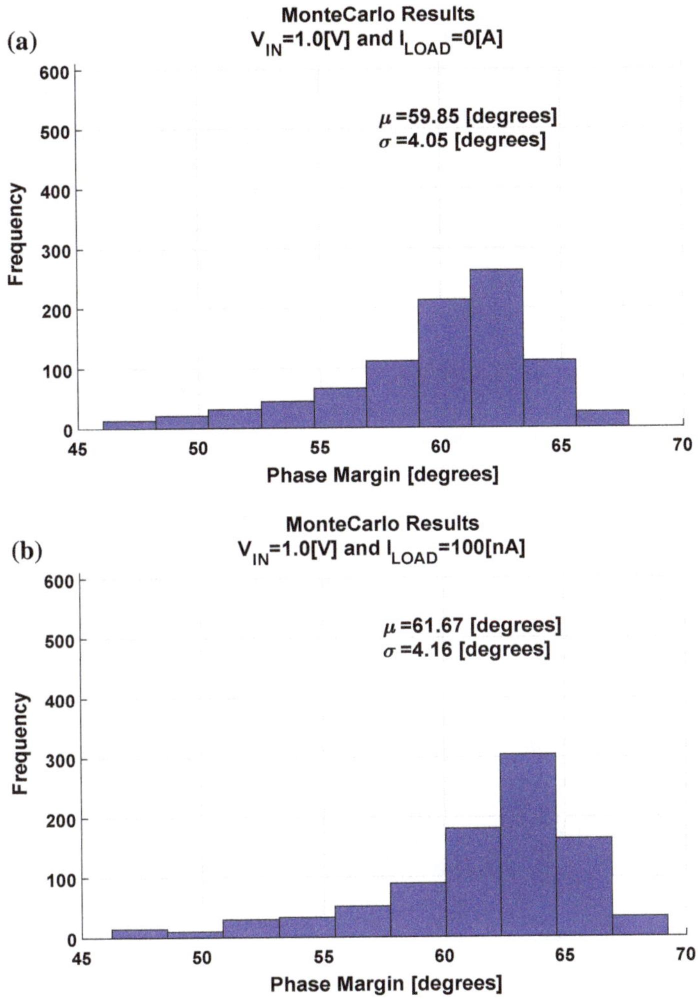

Fig. 3.18 Monte Carlo results for $V_{IN} = 1.0\,V$, $C_{LOAD} = 100\,pF$, and: **a** $I_{LOAD} = 0\,nA$, **b** $I_{LOAD} = 100\,nA$

transient response for $V_{IN} = 1.0\,V$, $C_{LOAD} = 100\,pF$, and rise/fall times of 100 ns and 1 μs, respectively. In both cases, the load current changes from its maximum, $I_{LOAD} = 100\,mA$, to its minimum values, $I_{LOAD} = 0\,A$. Measurements show a maximum overshoot of 0.242 V, and undershoot of 0.336 V, for a rise/fall time of 100 ns. The worst measured settling time is 17.05 μs.

The line transient response is shown in Fig. 3.24 for V_{IN} changing from 1.0 to 1.2 V, which is the maximum supply voltage admissible by the technology. The rise and fall times are 100 ns. Once again, the test setup has been configured for the worst-case scenario using $C_{LOAD} = 100\,pF$ and $I_{LOAD} = 100\,mA$. The voltage V_{OUT} remains

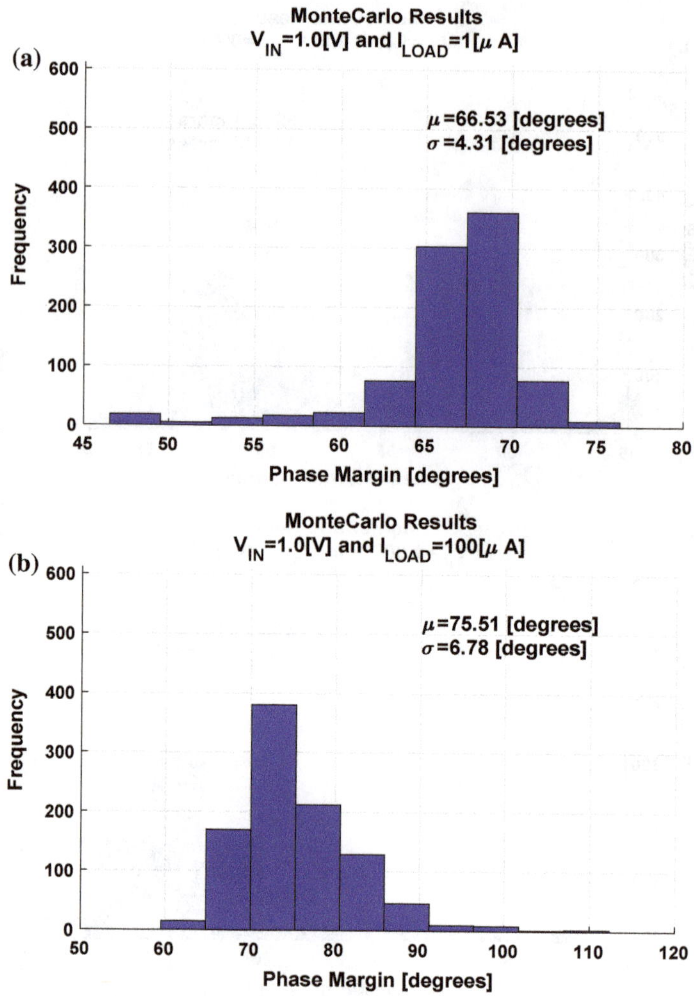

Fig. 3.19 Monte Carlo results for $V_{IN} = 1.0\,V$, $C_{LOAD} = 100\,pF$, and: **a** $I_{LOAD} = 1\,\mu A$, **b** $I_{LOAD} = 100\,\mu A$

within 80 mV with respect to its nominal value, and the worst measured settling time is 0.754 μs. Experimental results show that the proposed technique stabilizes the IC-LDO regulator in the whole current load range.

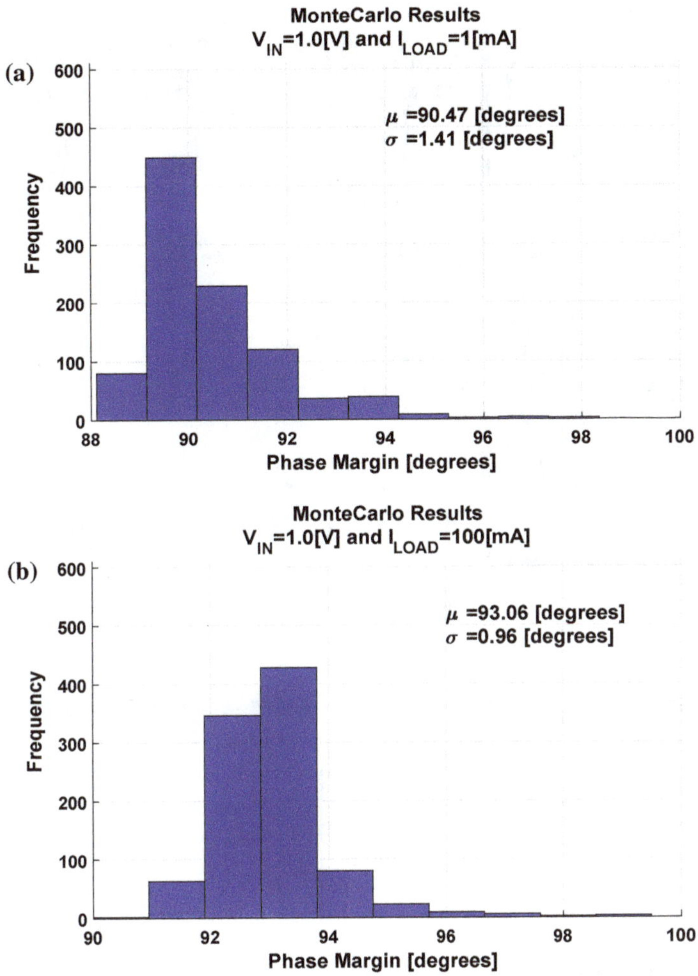

Fig. 3.20 Monte Carlo results for $V_{IN} = 1.0\,V$, $C_{LOAD} = 100\,pF$, and: **a** $I_{LOAD} = 1\,mA$, **b** $I_{LOAD} = 100\,mA$

3.6 Comparison with the State of the Art

Table 3.5 shows a comparison of the proposed LDO regulator with other recently published ones. In addition, the FOM_1, as defined in Eq. 2.24, has been computed for the proposed LDO regulator, and Fig. 3.25 shows a graphical comparison of the FOM_1 for the IC-LDO regulators of Table 3.5.

According to Table 3.5, [54] outperforms the rest of the structures in the comparison. In that paper, authors proposed an adaptive pass transistor to manage I_{LOAD}. Despite its small FOM_1, such a scheme is not very efficient for large values of I_{LOAD}.

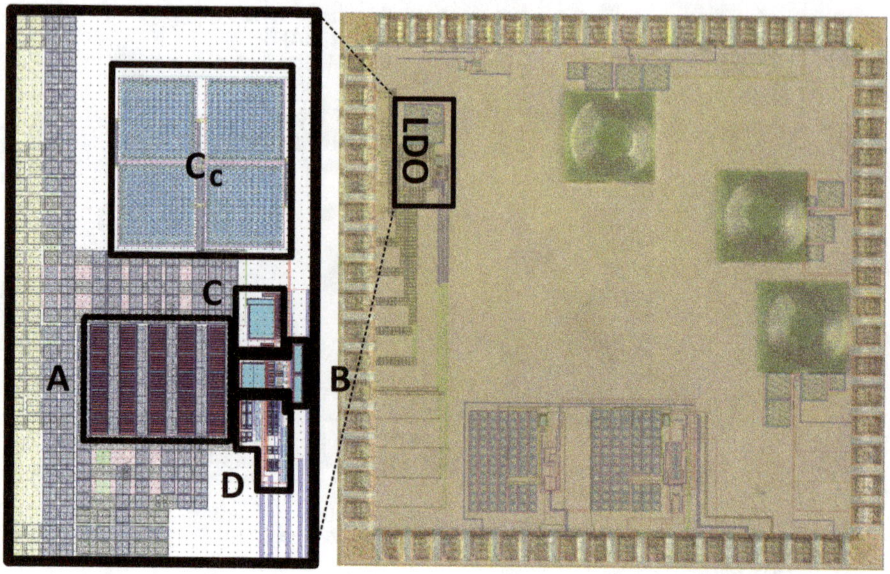

Fig. 3.21 IC-LDO regulator layout superimposed on a multi-project chip microphotograph

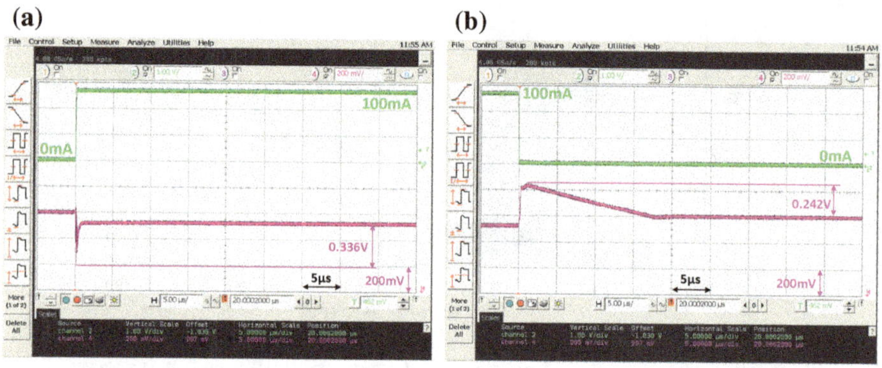

Fig. 3.22 Measured load transient response with $C_{LOAD} = 100\,pF$ and $V_{IN} = 1.0\,V$. For I_{LOAD} changing: **a** From 0 to 100 mA, and **b** from 100 to 0 mA, with rise and fall times of 100 ns

As a matter of fact, when the load current increases, a high gain, three-stage amplifier is needed to drive the main power transistor, resulting in a significant increase in the total quiescent power consumption, which increases the value of the FOM_1. Note that this increase in power consumption is not considered in Table 3.5. In the work presented here, the quiescent power consumption remains constant regardless the value of I_{LOAD}. Moreover, the small value of the FOM_1 in [54] is partly due to the addition of some specific fast path blocks, which were included to enhance the transient response. These blocks could have also been used in this regulator, but it is beyond our interest, which is the proposition of a new compensation method.

Table 3.5 Comparison of recently published LDO regulators

		[30]	[37]	[41][a]	[42]	[44]	[54]	[110]	This work	This work*
Process	[μm]	0.6	0.35	0.18	0.35	0.5	0.065	0.065	0.065	
V_{IN}	[V]	1.5–4.5	3.0–4.0	1.1–1.5	1.2–1.5	1.4–4.2	1.2	1.2	1.0–1.2	
V_{OUT}	[V]	1.3	2.8	1.0	1.0	1.21	1.0	1.0	0.8	
$V_{DROPOUT}$	[mV]	200	200	100	200	200	200	200	200	
$I_{LOAD,max}$	[mA]	100	50	50	50	100	100	50	100	
I_Q^{b}	[μA]	38	65	54	45	45	82.4	23.7	17.88	
C_{OUT}	[pF]	le2	le2	le2	le3	le5	le2	le4	100	
$\eta\|_{I_{LOAD,max}}$	[%]	99.962	99.935	99.946	99.955	99.955	99.917	99.953	99.982	
Area	[mm²]	0.307	0.350	–[c]	0.4	0.263	0.002	-[c]	0.0436	
Response time[b]	[μs]	2	15	2	4	5[d]	6	1.65	17.05	
ΔV_{OUT} varying V_{IN}										
• Maximum	[mV]	160	90	-[c]	-[c]	23	8.91	-[c]	78.4	
• Minimum	[mV]	−1.5	−10	-[c]	-[c]	-12	−10.63	-[c]	-49.3	
$\Delta V_{IN}/t_r^{e}$	[V/μs]	3/6	1/1	-[c]	-[c]	1/0.1	0.2/10	-[c]	0.2/0.1	
ΔV_{OUT} varying I_{LOAD}										
• Maximum	[mV]	100	80	100	70	47	0	19	242	96.39
• Minimum	[mV]	−90	−80	−80	−70	−48	−68.8	−58	−336	−39.36
$\Delta I_{LOAD}/t_r^{e}$	[mA/μs]	90/0.5	50/1	50/1	49/1	99.999/0.1	100/0.3	50/0.1	100/0.1	90/0.5
Line regulation	[mV/V]	-[c]	-[c]	-[c]	0.098	-[c]	4.7	8.89	26.5	
Load regulation	[μV/mA]	-[c]	-[c]	-[c]	250	408	300	34	780.57	
FOM_1	[fs]	72.2	416	388.8	2520	42750	56.69	7300	103.35	24.37

[a]Simulation results, [b]Worst Case, [c]Not available, [d]Estimation based on published results, [e]t_r: Rise time

*Performances and FOM_1 for this work under the conditions of [30]

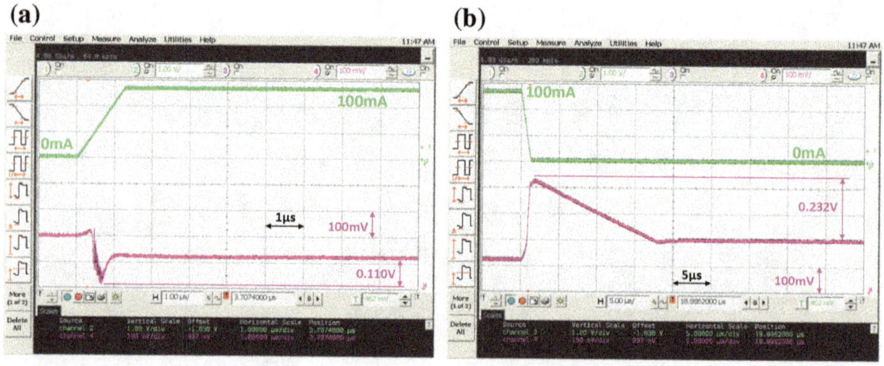

Fig. 3.23 Measured load transient response with $C_{LOAD} = 100\,pF$ and $V_{IN} = 1.0\,V$. For I_{LOAD} changing: **a** From 0 to 100 mA, and **b** from 100 to 0 mA, with rise and fall times of 1 µs

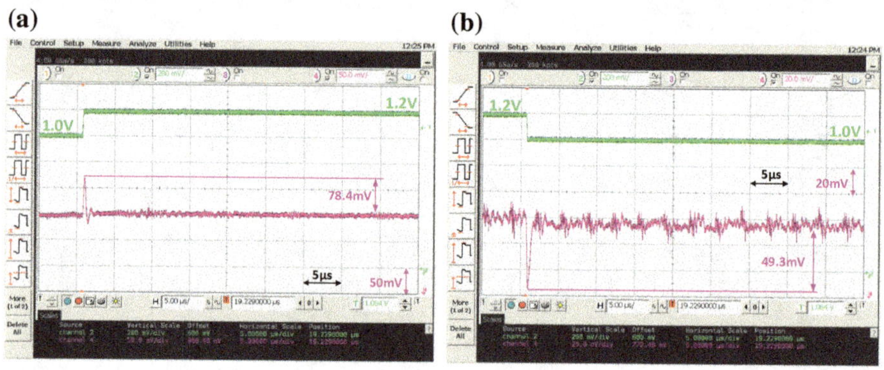

Fig. 3.24 Measured line transient response with $C_{LOAD} = 100\,pF$ and $I_{LOAD} = 100\,mA$. For V_{IN} changing: **a** From 1.0 to 1.2 V, and **b** from 1.2 to 1.0 V, with rise and fall times of 100 ns

The second best FOM_1 is achieved by [30]. However, its authors do not offer test results for I_{LOAD}, min $= 0\,A$. In order to get a fair comparison, a new FOM_1 value is estimated for this work under the same conditions than [30], that is, I_{LOAD} switching from 10 to 100 mA with a rise and fall times of 500 ns. Under this new condition, the regulator proposed here obtains a maximum variation of $+96.93\,mV/-39.36\,mV$ with respect to the nominal $V_{OUT} = 0.8\,V$, yielding a new value of FOM_1 of 24.37 fs, who is the best FOM_1 of Table 3.5.

In addition, [30] cannot handle a zero-load current. Figure 3.25 shows the FOM_1 for the IC-LDO regulators included in Table 3.5 versus the quiescent current consumption. Note that, FOM_1 for the proposed regulator has been re-computed under the conditions in [30], indicated as "This work*". From that figure, it can be concluded that the proposed LDO regulator (with the new compensation technique) is in the state of the art.

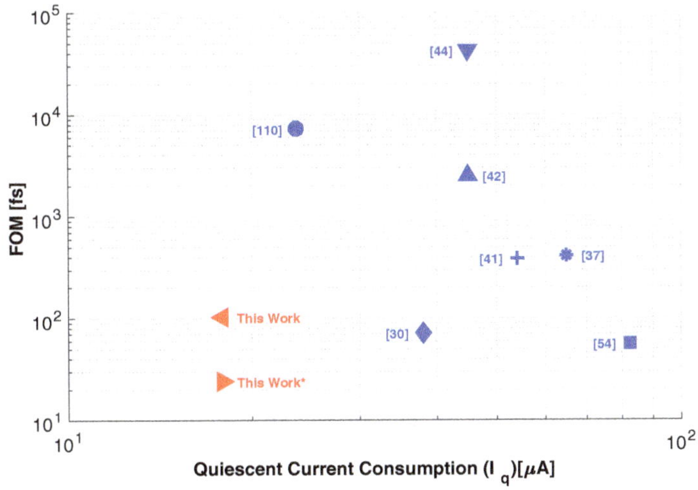

Fig. 3.25 Graphical comparison of the FOM_1 for the IC-LDO regulators of Table 3.5

3.7 Conclusions of This Chapter

In this chapter, an effective and simple frequency compensation technique for analogue cells which have to handle very large variations of I_{LOAD} is proposed. These variations produce a significant change in their frequency response, so that it is difficult to devise a compensation scheme valid for the whole range of load currents. In fact, the classical Miller compensation, with a fixed value of the RC compensating network, cannot ensure the stability in the whole range. The proposed adaptive nulling resistor technique proposed in this chapter for Miller compensation has proven to be a promising solution that achieves a good phase margin for the whole range of I_{LOAD}.

In order to validate the effectiveness of this technique, an IC-LDO regulator has been designed. It is a representative example, as these cells suffer not only from large variations of the load current but also from variations in the input voltage and in the output capacitance. When the proposed technique is particularized to IC-LDO regulators, it results in a simple, robust and power-efficient solution. This circuit has been designed and manufactured in a 65 nm standard CMOS technology. As it can be observed from Monte Carlo post-layout simulations and from experimental measurements, the proposed technique stabilizes the LDO regulator for the whole I_{LOAD} range, from 0 to 100 mA. The measured value of a typical FOM shows that it is in the state of the art. Even more, additional transient response enhancement techniques could also be applied to improve the transient response. Regarding current and area consumption, the circuitry required by the proposed compensation method only consumes 10% of the total quiescent power consumption and negligible area.

Chapter 4
Ultra-Low Quiescent Power Consumption LDO Regulators

Abstract This chapter discusses challenges introduced by ultra-low power consumption in power management circuits. First of all, some techniques proposed in the literature to design ultra-low-power LDO regulators are reviewed. Then, an IC-LDO regulator with a quiescent current consumption lower than 600 nA is proposed. It is based on the classical LDO topology, which has been modified to include a class AB buffer between the output of the error amplifier and the gate of M_{PASS}. This way, a fast charge/discharge of its parasitic capacitance is achieved with the inherent low quiescent power consumption of class AB circuits. The proposed regulator has been fabricated in a standard 0.18 μm CMOS technology. Experimental results show that the proposed regulator has a Figure of Merit in the state of the art.

4.1 Introduction

Power management systems have become very important in recent years, especially in the design of SoCs for low-power applications [111]. Of special interest is their use in Wireless Sensor Networks (WSNs) [112], where the sensor nodes are usually powered by batteries and in some cases, the energy they consume is harvested from the environment [113]. In this scenario, reducing the energy consumption of the blocks that build these nodes means prolonging the lifetime of the batteries, and/or reducing the requirements of the energy harvester.

When the main design requirement for an LDO regulator is ultra-low power consumption (for instance, a quiescent current lower than 1 μA), two critical aspects must be taken into account. First, extremely low bias currents make the output impedance of the error amplifier (as well as that of any intermediate stage) very high. This translates into low-frequency poles that degrade the frequency response and therefore, the stability. Compensation techniques based on active blocks, like those presented in [30, 36, 37] should be avoided due to the extra power consumption they consume. Second, extremely low quiescent currents slow down the transient response, as they are not sufficient to rapidly charge/discharge the high gate capacitance of the pass transistor.

© Springer International Publishing AG, part of Springer Nature 2018

J. M. Hinojo et al., *Internally Compensated LDO Regulators for Modern System-on-Chip Design*, https://doi.org/10.1007/978-3-319-75411-6_4

To improve the stability and transient response, a low quiescent power buffer placed between the error amplifier and the pass transistor is appropriate. To avoid an increase in power consumption, a class AB buffer is the right choice.

4.2 Ultra-Low-Power Design Challenges

In the last years, power consumption has become one of the most important concerns for the scientific community, as a consequence of the massive impact of WSNs, which are part of the so-called Internet of Things. Among the most popular IoT applications we can find smart metering [114–116], biomedical and implantable sensing [117–121], structural health [122–124] and environmental monitoring [116, 125–127].

Depending on the application, the number of sensor nodes in a WSN varies from tens to thousands. Therefore, designers try to miniaturize sensor nodes in order to reduce their impact on the surrounding environment and cost [128]. This miniaturizing trend also affects the battery size, which is proportional to the energy they are able to store [125]. In this context, the battery lifetime can be computed by means of the expression:

$$T_{LF} \approx \frac{E_{battery}}{P_{avg}} \tag{4.1}$$

where T_{LF} represents the battery lifetime, $E_{battery}$ is the capacity of the battery (meaning the total energy that the battery can provide to the node) and P_{avg} is the average power consumption of the sensor node. Based on Eq. 4.1, for a battery with a capacity of 0.1 mWh, extending its lifetime up to 10 years, which is an adequate value for relevant applications [117], constrains the average power consumption to the order of nanowatts or less.

In order to reduce P_{avg}, it is necessary to analyse the sensor node operation. In general, most of WSN-based applications perform periodic tasks. After a certain active time (T_{duty}), whose duration is smaller than the period (T_{CYCLE}), the sensor node completes its tasks. The rest of the period, $T_{sleep} = T_{CYCLE} - T_{duty}$, the node is in sleep mode drastically reducing its power consumption. Figure 4.1 shows a common use case for a sensor node. During the time interval I_0, the sensor node is actively

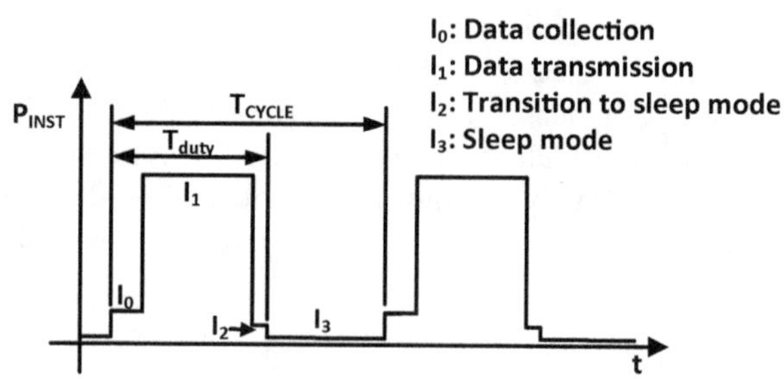

Fig. 4.1 Typical power consumption pattern of a sensor node in a WSN

collecting data from the surrounding environment. Next, during interval I_1, data is transmitted to a central node that stores, processes and sends the useful information to its final destination. Once this transmission ends, the sensor node prepares itself to go into sleep mode, during interval I_2. Finally, the sensor node remains in sleep mode, during interval I_3, until a new cycle begins. Assuming this pattern, there are three major contributions to P_{avg}:

- **Duty-Cycle power, P_{duty}**: accounts for the average power consumption of those subsystems in the node which are active during the duty cycle, excluding the quiescent power consumption,
- **Sleep mode power, P_{sleep}**: which is given by the average power consumption of those blocks which are active during the sleep mode (such as the sleep time counters), excluding the quiescent power consumption, and
- **Quiescent power, $P_{quiescent}$**: that takes into account the quiescent power consumption of those blocks (e.g. the main clock circuitry) that must remain active during the entire cycle.

Therefore, P_{avg} can be expressed as

$$P_{avg} = P_{duty} + P_{sleep} + P_{quiescent} \tag{4.2}$$

where the duty-cycle power, P_{duty}, and sleep power, P_{sleep}, can be estimated as

$$P_{duty} = \frac{1}{T_{CYCLE}} \int_0^{T_{DUTY}} P_{INST}(t)\,dt = \frac{1}{T_{CYCLE}} \sum_{k=0}^{N-1} P_k T_k = \frac{E_{duty}}{T_{CYCLE}} \tag{4.3}$$

$$P_{sleep} = \frac{P_3 \cdot T_3}{T_{CYCLE}} = \frac{E_{sleep}}{T_{CYCLE}} \tag{4.4}$$

In Eqs. 4.3 and 4.4, $P_{INST}(t)$ is the instantaneous power consumed by the sensor node in time t (excluding the quiescent power consumption), N is the total number of phases required to collect data and transmit them to the central node and P_K and T_K represent the power consumed and the time spent, respectively, in completing the k-th phase, $\sum_{k=0}^{N-1} T_k = T_{duty}$. P_3 is the average power consumption during the sleep mode, excluding the quiescent power consumption.

In the same way, E_{duty} and E_{sleep} are the energy consumed during the duty cycle and the sleep mode, respectively, excluding the quiescent consumption. Consequently, the average power can be expressed as

$$P_{avg} = P_{duty} + P_{sleep} + P_{quiescent} = \frac{E_{duty}}{T_{CYCLE}} + P_{sleep} + P_{quiescent} \tag{4.5}$$

The above analysis is not directly applicable to cases where the sensor nodes include energy scavenging devices. Here, the function of the battery is to ensure the operation of a node when the external source of energy is not available or does not provide sufficient energy. In those situations, the lifetime of the sensor is normally determined by the availability of the external source, rather than by the capacity of the battery [129].

There is a recent trend towards the design of battery-less sensor nodes, which only operate when the energy scavenging device provides enough instantaneous energy to supply the node. In this case, the critical parameter is the instantaneous power consumed by the sensor node, which must be smaller, at all times of T_{duty}, than that provided by the energy scavenging device [130].

For battery-powered devices, the most straightforward way to reduce P_{avg} is to increase the sleep time by increasing the period T_{CYCLE}. This technique is called duty cycling. However, this solution reduces the acting and monitoring capability of the WSN, as it decreases the number of completed tasks in a given time interval, e.g. in one day. As T_{CYCLE} cannot be indefinitely extended, in order to further decrease P_{avg}, complementary techniques to reduce P_{duty}, such as clock and power gating, Dynamic Frequency and Voltage Scaling (DFVS) or multiple voltage domains, have been also used [131].

Regarding $P_{quiescent}$, it is not affected by duty cycling. Thus, the only way to reduce $P_{quiescent}$ is to design the circuits contributing to this term with less quiescent power consumption. As a sensor node requires a well-regulated supply voltage to properly operate, even in sleep mode, the power management subsystem is one of the critical contributors to the quiescent power consumption. Consequently, researchers' interest has been recently focused on reducing the static power consumption of these circuits as much as possible [132–144].

Within the power management circuits, one of the most critical blocks is the voltage regulator, as it supplies sensitive subsystems such as RF stages or analogue-to-digital converters. One example of an IC-LDO regulator whose quiescent power consumption is lower than 1 μW is proposed in [54]. The authors split the pass transistor into two smaller transistors in order to minimize the quiescent current consumption of the error amplifier for low I_{LOAD}, as it was described in Sect. 2.2.2. Another example can be found in [47], where the authors propose a high slew-rate implementation of the error amplifier to avoid the limitations that a low quiescent current consumption imposes on the transient response. It uses an input stage based on two common-gate amplifiers that can supply a current higher than its bias current.

In this chapter, a new IC-LDO regulator with ultra-low quiescent power consumption is presented. Based on a classical LDO topology, it includes, at the output of the error amplifier, a class AB buffer, which is able to provide large charging/discharging current to the parasitic capacitance of M_{PASS}. This buffer improves the transient response of the proposed IC-LDO regulator with an ultra-low quiescent current consumption.

4.3 Design of an Ultra-Low-Power IC-LDO Regulator

The proposed LDO regulator is shown in Fig. 4.2. As it can be seen in Fig. 4.3a, the error amplifier consists of a Folded Cascode operational amplifier, including the inbuilt A_0 and A_1 amplifiers. They are low voltage, high signal range amplifiers [145] that improve the open-loop gain of the whole system. For the sake of completeness, their schematics are reproduced in Fig. 4.3b and c, respectively.

To enlarge the signal range and increase the slew-rate at the gate of M_{PASS}, a class AB voltage buffer [146] has been placed at the output of the error amplifier. Class AB operation is achieved by means of RC coupling, as depicted in Fig. 4.2b.

Under steady-state conditions, the resistor R_{BUF} sets the gate voltage, V_Y, of transistor M_{16} to $V_{BP,2}$, which determines the bias current of the buffer. Under transient conditions, the variations of the voltage V_X at the drain of M_{15} (which, in turn, is the gate of M_{14}) are immediately transferred to the gate of M_{16} through the coupling capacitor C_{BUF}, so that the instantaneous output current of the buffer increases or decreases according to the needs of the load. As an example, assuming that the voltage at the input of the buffer $v_{IN,B}$ decreases, then the voltage at the drain of M_{15} (gate of M_{14} or V_X) increases, and, due to the coupling capacitor C_{BUF}, the voltage at the gate of M_{16}, V_Y, increases as well; this reduces the current through M_{16} and increases that of M_{14}, causing a fast discharge of the parasitic capacitance at the gate of M_{PASS} (C_{GATE}, which is also the output of the buffer. The opposite effect occurs when the voltage at the input of the buffer increases, which causes a significant increase of the current of transistor M_{16}, and a decrease that of M_{14}, which leads to a fast charge of C_{GATE}. The transient effect vanishes when the current flowing through the resistor R_{BUF} takes the voltage at the coupling capacitor C_{BUF} to its final value.

The proposed regulator has been designed and manufactured in a standard 0.18 μm CMOS technology for a nominal output voltage of 1 V, a maximum I_{LOAD} of 100 mA and V_{REF} of 0.8 V. Transistor aspect ratios and the value of the most significant passive components are summarized in Tables 4.1 and 4.2, respectively.

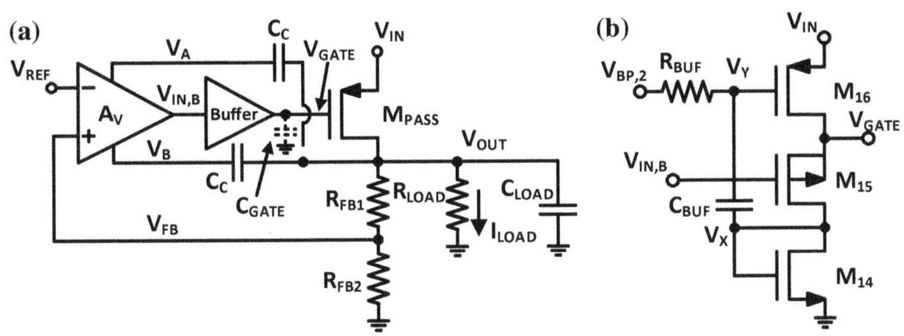

Fig. 4.2 **a** Simplified view of the proposed ultra-low-power IC-LDO regulator. **b** Schematic of the class AB buffer driving M_{PASS}

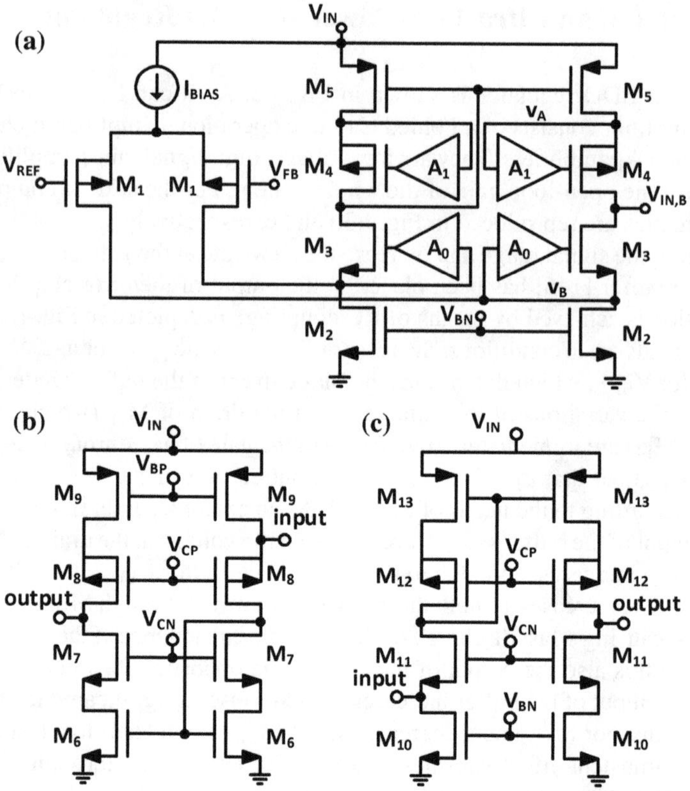

Fig. 4.3 Additional blocks included in the proposed IC-LDO regulator: **a** Schematic of the error amplifier, and a detailed view of the inbuilt amplifiers: **b** A_1, **c** A_O

Table 4.1 Transistor aspect ratios

Devices	Aspect ratios [μm/μm]	Devices	Aspect ratios [μm/μm]
M_{PASS}	12500/0.18	M_1	10/0.36
$M_2 - M_3$	4/0.36	$M_4 - M_5$	20/0.36
$M_6 - M_7$; $M_{10} - M_{11}$	2/3.6	$M_8 - M_9$; $M_{12} - M_{13}$	10/3.6
M_{14}	4/0.36	M_{15}	10/0.36
M_{16}	40/0.36		

Table 4.2 Main passive element values

Passive component	Value
R_{FB1}	12.5 [MΩ]
R_{FB2}	60.0 [MΩ]
R_{BUF}	50.0 [MΩ]
C_{BUF}	4.0 [pF]
C_C	0.8 [pF]

4.3.1 Stability Analysis

The stability of the proposed IC-LDO regulator is analysed using the small-signal equivalent model, represented in Fig. 4.4. The open-loop transfer function is shown in Eq. 4.6, where, as usual, $g_{m,i}$ and $r_{o,i}$ represent the transconductance and output resistance of the *i*-th transistor, respectively. The capacitance $C_{OUT,EA}$ models the lumped capacitance at the output of the error amplifier. It should be noted that the passive elements C_{BUF} and R_{BUF} have been included in this analysis to evaluate their impact on Eq. 4.6. Capacitors C_C are included to compensate the proposed regulator, according to the Hybrid Cascode Feedforward Compensation (HCFC) [102, 147–150]. In addition, in Eqs. 4.10–4.19, $R_{GB,NMOS}$ and $R_{GB,PMOS}$ represent the equivalent output resistance of amplifiers A_0 and A_1, respectively. $C_{gg,NMOS}$ and $C_{gg,PMOS}$ model the parasitic capacitance at the gate of M_3 and M_4. Finally, R_1 represents the output resistance of M_{14}, which is given by $R_1 = r_{o,14} \| 1/g_{m,14}$.

$$H(s) = A_{OL}\frac{\left(1+\frac{s}{a_1}+\frac{s^2}{a_2}\right)\left(1+\frac{s}{b_1}+\frac{s^2}{b_2}+\frac{s^3}{b_3}\right)}{\left(1+\frac{s}{\omega_{p1}}\right)\left(1+\frac{s}{\omega_{p2}}\right)\left(1+\frac{s}{\omega_{p3}}\right)\left(1+\frac{s}{c_1}+\frac{s^2}{c_2}\right)\left(1+\frac{s}{\omega_{p6}}\right)} \tag{4.6}$$

The output resistance, open-loop gain and the dominant pole are given by Eqs. 4.7–4.9, respectively.

$$R_{OUT} = r_{o,P} \| R_{LOAD} \| \left(R_{FB1} + R_{FB2}\right) \tag{4.7}$$

$$A_{OL} = g_{m,1}\left[\left[A_0 g_{m,3} r_{o,3}(r_{o,1} \| r_{o,2})\right] \| \left(A_1 g_{m,4} r_{o,4} r_{o,5}\right)\right] g_{m,P} R_{OUT} \tag{4.8}$$

$$\omega_{p1} = \frac{1}{2 g_{m,P} R_{OUT}\left[\left[A_0 g_{m,3} r_{o,3}\left(r_{o,1} \| r_{o,2}\right)\right] \| \left(A_1 g_{m,4} r_{o,4} r_{o,5}\right)\right] C_C} \tag{4.9}$$

The non-dominant contribution of the denominator of Eq. 4.6 is composed of three real poles and a pair of complex poles, whose values are expressed in Eqs. 4.10–4.14. The zeros of the transfer function can be determined from Eqs. 4.15–4.19.

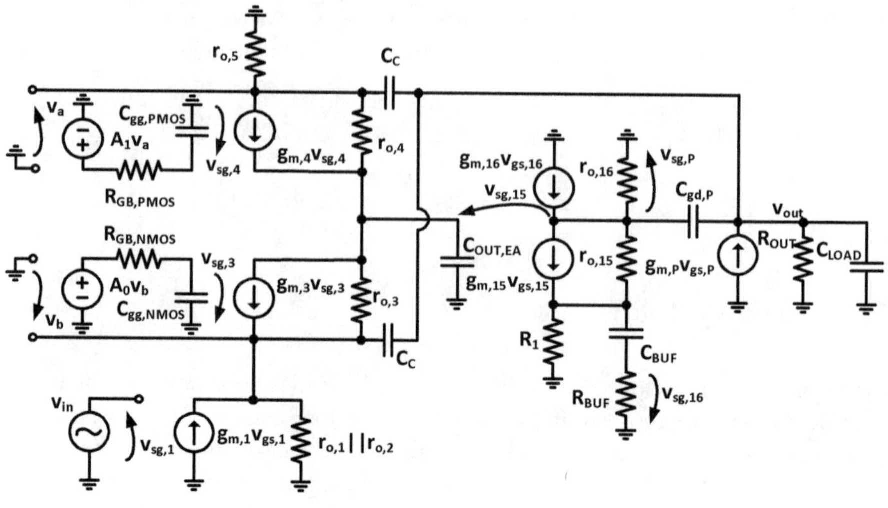

Fig. 4.4 Small-signal model of the proposed IC-LDO regulator

$$\omega_{p2} = \frac{2g_{m,3}\left(A_0R_{GB,PMOS}C_{gg,PMOS} + A_1R_{GB,NMOS}C_{gg,NMOS}\right)}{R_{GB,NMOS}C_{gg,NMOS}\left(A_1C_C + 2g_{m,3}R_{GB,PMOS}C_{gg,PMOS}\right)} \tag{4.10}$$

$$\omega_{p3} = \frac{g_{m,4}g_{m,15}r_{o,3}\left(1+g_{m,16}R_1\right)\left(A_1C_C + 2g_{m,3}R_{GB,PMOS}C_{gg,PMOS}\right)C_C}{R_{GB,PMOS}C_{gg,PMOS}\left[g_{m,4}C_{gd,P}\left(C_C + g_{m,3}r_{o,3}C_{OUT,EA}\right) + g_{m,15}r_{o,3}\left(g_{m,3}+g_{m,4}\right)\left(1+g_{m,16}R_1\right)C_C^2\right]} \tag{4.11}$$

$$c_1 = \frac{r_{o,4}\left[g_{m,4}C_{gd,P}\left(C_C + g_{m,3}r_{o,3}C_{OUT,EA}\right) + g_{m,15}\left(g_{m,3}+g_{m,4}\right)r_{o,3}\left(1+g_{m,16}R_1\right)C_C^2\right]}{C_CC_{gd,P}\left[C_C\left(r_{o,3}+r_{o,4}\right) + C_{OUT,EA}\left(g_{m,3}+g_{m,4}\right)r_{o,3}r_{o,4}\right]} \tag{4.12}$$

$$c_2 = \frac{g_{m,P}R_{OUT}R_{BUF}\left[C_{gd,P}g_{m,4}\left(C_C + C_{OUT,EA}g_{m,3}r_{o,3}\right) + C_C^2\left(g_{m,3}+g_{m,4}\right)r_{o,3}g_{m,15}\left(1+g_{m,16}R_1\right)\right]}{C_CC_{gd,P}C_{OUT,EA}r_{o,3}\left[C_CR_{BUF}\left(1+g_{m,P}R_{OUT}\right) + C_{LOAD}R_{BUF}R_{OUT}g_{m,3} + C_Cg_{m,P}R_1R_{OUT}\right]} \tag{4.13}$$

$$\omega_{p6} = \frac{C_CR_{BUF}\left(1+g_{m,P}R_{OUT}\right) + C_{LOAD}R_{BUF}R_{OUT}g_{m,3} + C_Cg_{m,P}R_1R_{OUT}}{R_{BUF}R_{OUT}C_{LOAD}C_C} \tag{4.14}$$

$$a_1 = \frac{g_{m,4}\left(A_0R_{GB,PMOS}C_{gg,PMOS} + A_1R_{GB,NMOS}C_{gg,NMOS}\right)}{R_{GB,PMOS}C_{gg,PMOS}\left(A_0C_C + g_{m,4}R_{GB,NMOS}C_{gg,NMOS}\right)} \tag{4.15}$$

$$a_2 = \frac{g_{m,4}\left(A_0R_{GB,PMOS}C_{gg,PMOS} + A_1R_{GB,NMOS}C_{gg,NMOS}\right)}{R_{GB,PMOS}C_{gg,PMOS}R_{GB,NMOS}C_{gg,NMOS}C_C} \tag{4.16}$$

$$b_1 = -\frac{g_{m,P}}{C_{gd,P}} \tag{4.17}$$

$$b_2 = -\frac{g_{m,3}g_{m,15}g_{m,P}R_{BUF}r_{o,6}r_{o,8}\left(1+g_{m,16}R_1\right)}{\alpha C_CC_{gd,P} + \beta C_CC_{OUT,EA} + \delta C_{gd,P}C_{OUT,EA}} \tag{4.18a}$$

$$\alpha = R_{BUF}\left(r_{o,3}+r_{o,4}\right) \tag{4.18b}$$

$$\beta = g_{m,15}R_{BUF}r_{o,3}r_{o,4}\left(1+g_{m,16}R_1\right) \tag{4.18c}$$

$$\delta = g_{m,4}r_{o,3}r_{o,4}\left(R_{BUF}+R_1\right) \tag{4.18d}$$

$$b_3 = -\frac{g_{m,3}g_{m,15}g_{m,P}\left(1 + g_{m,16}R_1\right)}{C_C C_{gd,P} C_{OUT,EA}} \tag{4.19}$$

Assuming that the second-order polynomials defined by c_1 and c_2, and a_1 and a_2, are the dominant contributors to the phase margin. Their damping coefficients $\xi_{c1,c2}$ and $\xi_{a1,a2}$ can be expressed by

$$\xi_{c_1,c_2} = \frac{r_{o,3}C_C C_{gd,P}C_{OUT,EA}\left(R_{BUF}\left[C_C\left(1 + g_{m,P}R_{OUT}\right)\right] + g_{m,P}R_{OUT}R_1C_C\right)}{2g_{m,P}R_{OUT}R_{BUF}\beta\sqrt{\frac{C_C C_{gd,P}\left[C_C\left(r_{o,3}+r_{o,4}\right)+C_{OUT,EA}\left(g_{m,3}+g_{m,4}\right)r_{o,3}r_{o,4}\right]}{r_{o,4}\beta}}} \tag{4.20}$$

$$\xi_{a_1,a_2} = \frac{R_{GB,PMOS}R_{GB,NMOS}C_{gg,PMOS}C_{gg,NMOS}C_C}{2g_{m,4}\theta\sqrt{\frac{R_{GB,PMOS}C_{gg,PMOS}\left(A_0C_C+g_{m,4}R_{GB,NMOS}C_{gg,NMOS}\right)}{g_{m,4}\theta}}} \tag{4.21}$$

where

$$\beta = g_{m,4}C_{gd,P}\left(C_C + g_{m,3}r_{o,3}C_{OUT,EA}\right) + g_{m,15}r_{o,3}\left(g_{m,3} + g_{m,4}\right)\cdot \cdot\left(1 + g_{m,16}R_1\right)C_C^2 \tag{4.22}$$

$$\theta = A_0R_{GB,PMOS}C_{gg,PMOS} + A_1R_{GB,NMOS}C_{gg,NMOS} \tag{4.23}$$

According to Eqs. 4.20 and 4.21, a value of R_{BUF} and C_C can be chosen to achieve a suitable frequency response peaking.

Figure 4.5 shows the Bode diagram for different values of I_{LOAD}, obtained through post-layout simulations. In all cases, the load capacitance is equal to 100 pF that, regarding stability, corresponds to the worst case. Table 4.3 shows the gain and phase margin obtained for different load currents. Figure 4.6 represents a simplified pole-zero diagram to show how their locations are modified when I_{LOAD} changes.

4.3.2 Experimental Results

The layout superimposed on a micrograph of the chip is shown in Fig. 4.7a, where the designed regulator has been remarked. Figure 4.7b presents a detailed view of the cell layout. Block *A* corresponds to M_{PASS}, block *B* is the error amplifier, block *C* points to the location of the buffer used to drive the gate of M_{PASS} and block *D* corresponds to the biasing circuitry, including the generation of the cascode voltages. In addition, the rectangle labelled as *E* contains decoupling capacitors required to stabilize cascode voltages. The total area occupied by the regulator, without the decoupling capacitors, is 315 × 460 μm, which corresponds to an area of 0.1449 mm^2. If decoupling capacitors are taken into account, the total area is 0.195 mm^2.

Experimental results have been obtained using the test setup of Appendix B. Figure 4.8a and b show the static response of the IC-LDO regulator for I_{LOAD} = 100 mA and I_{LOAD} = 0.1 mA, respectively. In both situations, C_{LOAD} is equal to

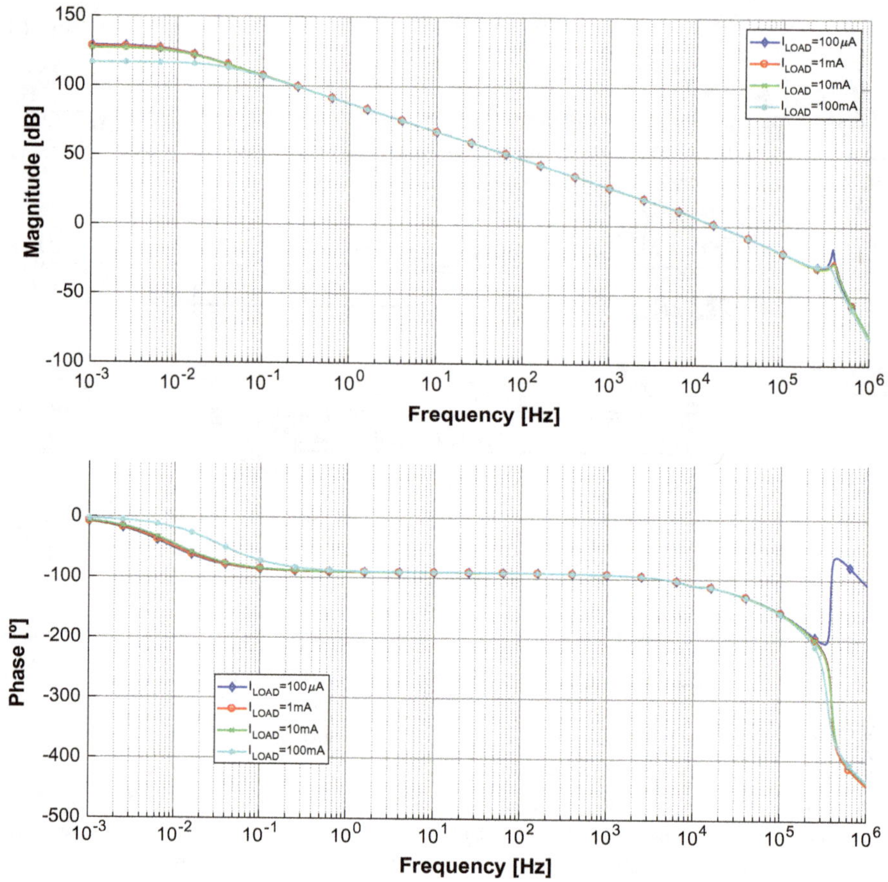

Fig. 4.5 Bode diagram for different values of I_{LOAD}

Table 4.3 Phase margin and DC gain of the transfer function of the proposed LDO regulator

I_{LOAD}	Gain $[dB]$	PM [°]
100[µA]	64.25	129.14
1[mA]	65.34	128.72
10[mA]	65.34	127.44
100[mA]	65.1	117.04

100 pF that, as stated before, is the worst-case condition regarding stability. To characterize the static behaviour of the regulator, a 10 ms period triangular signal is used as the input stimulus with an amplitude of 1.8 V_{pp}, and a 0.9 V DC level. As it can be seen in these figures, $V_{OUT} = 0.982$ V for $V_{IN} > 1.146$ V, which means a dropout voltage $V_{DROPOUT} = 0.164$ V.

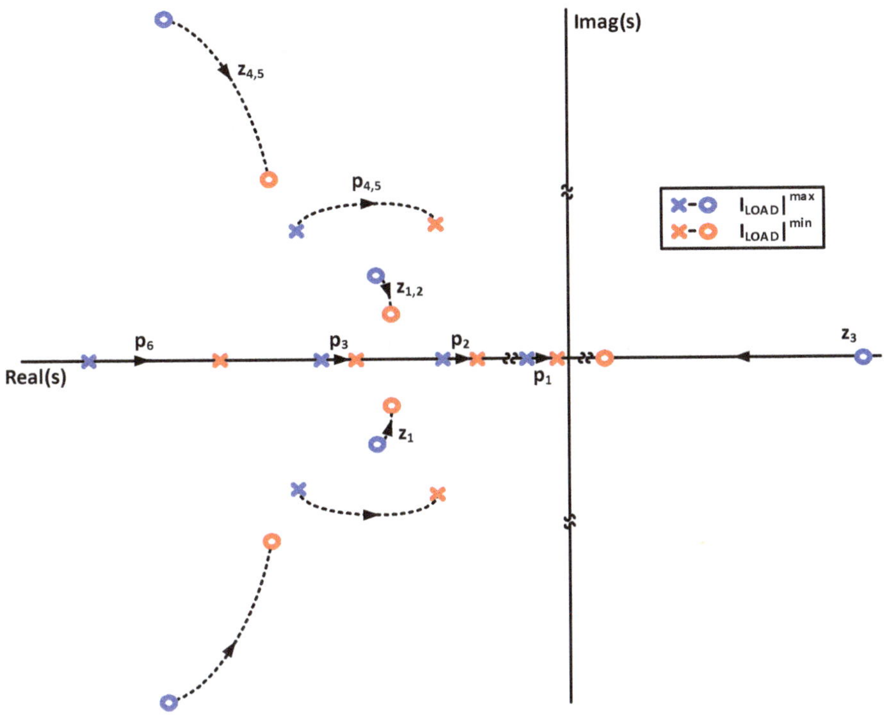

Fig. 4.6 Pole-zero diagram for different values of I_{LOAD}

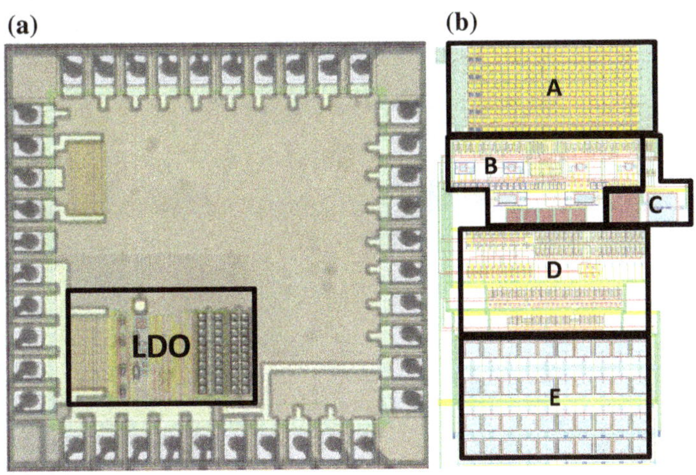

Fig. 4.7 **a** IC-LDO regulator layout superimposed on a multi-project chip micrograph. **b** A detail of the IC-LDO regulator layout

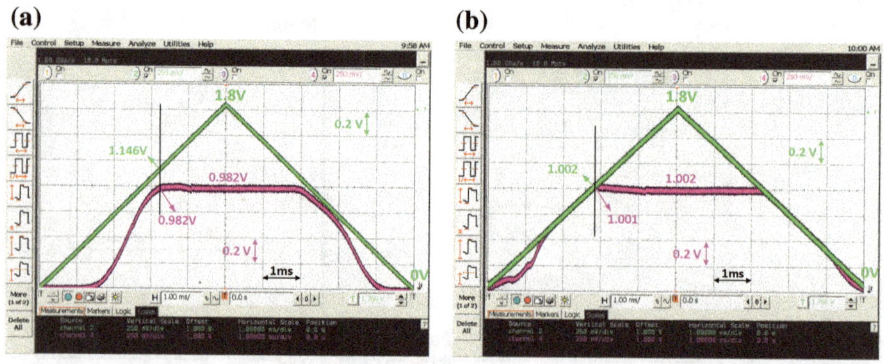

Fig. 4.8 Static response of the proposed regulator for: **a** $C_{LOAD} = 100\,pF$ and $I_{LOAD} = 100\,mA$. **b** $C_{LOAD} = 100\,pF$ and $I_{LOAD} = 100\,\mu A$

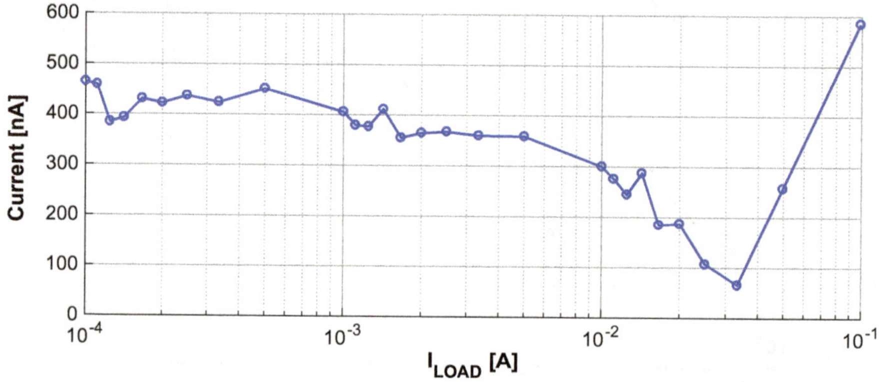

Fig. 4.9 Quiescent current variation of the proposed IC-LDO regulator for different I_{LOAD}

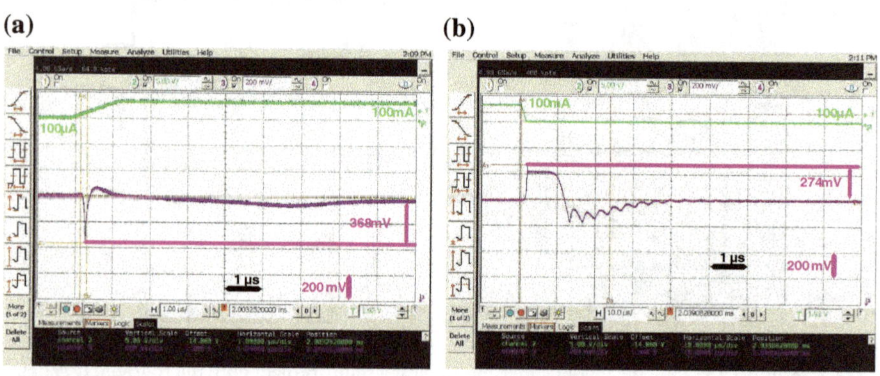

Fig. 4.10 Measured load transient response with $C_{LOAD} = 100\,pF$ and $V_{IN} = 1.2\,V$. For I_{LOAD} changing: **a** from 0.1 to 100 mA, and **b** from 100 to 0.1 mA, with rise and fall times of 1 μs

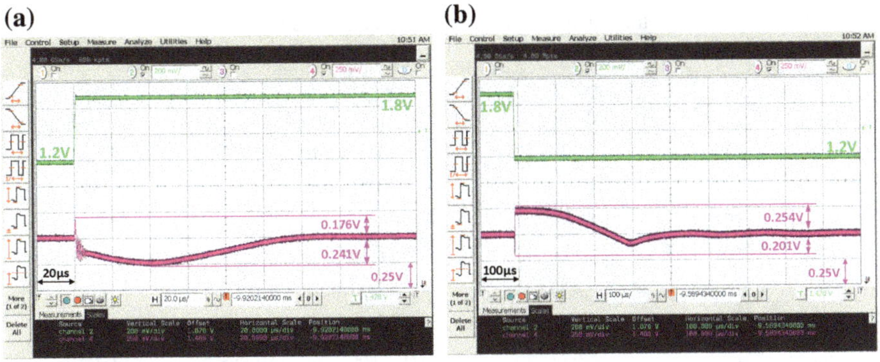

Fig. 4.11 Measured line response with $C_{LOAD} = 100\,\mathrm{pF}$ and $I_{LOAD} = 100\,\mu\mathrm{A}$. For V_{IN} changing: **a** from 1.2 to 1.8 V, and **b** from 1.8 to 1.2 V, with rise and fall times of 1 µs

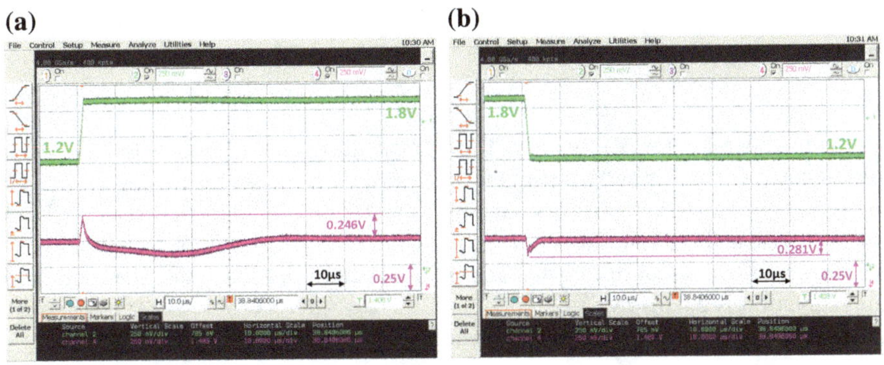

Fig. 4.12 Measured line response with $C_{LOAD} = 100\,\mathrm{pF}$ and $I_{LOAD} = 100\,\mathrm{mA}$. For V_{IN} changing: **a** from 1.2 to 1.8 V, and **b** from 1.8 to 1.2 V, with rise and fall times of 1 µs

Figure 4.9 shows the quiescent consumption of the proposed IC-LDO regulator for different load current values. They have been obtained using a potentiometer as the resistor load, whose value was changed to cover the entire operating range. The input voltage was set to its minimum value, $V_{IN} = 1.2\,\mathrm{V}$. Once again, $C_{LOAD} = 100\,\mathrm{pF}$.

Figure 4.10 shows the transient response of the regulator to variations in I_{LOAD} for $V_{IN} = 1.2\,\mathrm{V}$ and $C_{LOAD} = 100\,\mathrm{pF}$. Under these conditions, Fig. 4.10a and b show the variation of the output voltage when I_{LOAD} varies between $I_{LOAD,min} = 100\,\mu\mathrm{A}$ to $I_{LOAD,max} = 100\,\mathrm{mA}$ with a rise (t_r), and fall time (t_f), of 1 µs. In this situation, the variations of V_{OUT} are +274/−368 mV, with respect to its nominal value. The worst setup time was 23.58 µs. The measured load regulation is 85.44 µV/mA.

Figure 4.11a and b show the line response, voltage spikes at V_{OUT}, when V_{IN} changes from 1.2 and 1.8 V and viceversa, for $C_{LOAD} = 100\,\mathrm{pF}$ and $I_{LOAD} = 100\,\mu\mathrm{A}$. As in the previous case, rise and fall times were equal to 1 µs. Under these conditions, the maximum measured overshoot was 254 mV and the

Table 4.4 A summary of the performances of the proposed LDO regulator

		Proposed LDO			Proposed LDO
Process	[μm]	0.18	ΔV_{OUT} varying V_{IN}		
V_{IN}	[V]	1.164–1.8	• Maximum	[mV]	246
V_{OUT}	[V]	1.0	• Minimum	[mV]	−281
$V_{DROPOUT}$	[mV]	164	$\Delta V_{IN}/t_r^c$	[V/μs]	0.6/1
$I_{LOAD,max}$	[mA]	100	ΔV_{OUT} varying I_{LOAD}		
I_Q[a]	[μA]	0.585[b]	• Maximum	[mV]	274
C_{OUT}	[pF]	le2	• Minimum	[mV]	−368
$\eta\vert_{I_{LOAD,max}}$	[%]	99.9994	$\Delta I_{LOAD}/t_r{}^c$	[mA/μs]	99.9/1
Area	[mm^2]	0.195	Line regulation	[mV/V]	2.5
Response time[a]	[μs]	731.6	Load regulation	[μV/mA]	85.44
			FOM_1	[fs]	3.76

[a]Worst case, [b]Including the consumption of the biasing circuitry, [c]t_r: Rise time

minimum undershoot was −241 mV. The worst settling time was equal to 731.16 μs. Additionally, Fig. 4.12a and b depict the line transient response for $C_{LOAD} = 100\,pF$ and $I_{LOAD} = 100\,mA$, with V_{IN} changing from 1.2 to 1.8 V and vice versa. In both cases, the rise and fall times of V_{IN} are 1 μs. Under these conditions, the output voltage shows an overshoot of 246 mV and an undershoot of −281 mV. The measured line regulation is 2.5 mV/V.

Table 4.4 resumes the main performances of the proposed regulator.

4.4 Comparison with the State of the Art

Table 4.5 shows a comparison of the performances of the proposed regulator versus those that obtain the best Figure of Merit (FOM_1). In addition, Table 4.5 shows the values of their corresponding Figures of Merit (FOM_1), as defined in Eq. 2.24 (Chap. 2, Sect. 2.3). Figure 4.13 depicts the FOM_1 for the set of IC-LDO regulators versus their quiescent current consumption. The closer to the origin of coordinates, the better the regulator.

From Table 4.5 and Fig. 4.13, it can be concluded that the proposed LDO regulator clearly outperforms the rest of regulators considered in this comparison, as it achieves the best FOM_1 with the lowest quiescent current consumption.

4.5 Conclusions of This Chapter

One of the main requirements imposed by recent trends in microelectronics is low power consumption. In this sense, ultra-low-power voltage regulators are critical blocks to maximize the lifetime of the batteries and to enable autonomous, battery-

Table 4.5 Comparison of low-power IC-LDO regulators based on the classical topology

		[47]	[53]	[54]	[55]	[56]	[80]	This work
Process	[μm]	0.18	0.35	0.065	0.35	0.11	0.13	0.18
V_{IN}	[V]	1.0–1.8	2.5–4.0	1.2	1.2	1.8–3.8	1.15–1.4	1.164–1.8
V_{OUT}	[V]	0.9	2.35	1.0	1.0	1.6–3.6	1.0	1.0
$V_{DROPOUT}$	[mV]	100	150	200	200	200	150	164
$I_{LOAD,max}$	[mA]	50	100	100	100	200	50	100
I_q^a	[μA]	1.2	7.0	82.4	14.0	41.5	37.0	0.585
C_{OUT}	[pF]	1e2	1e2	1e2	1e2	4e1	2e1	1e2
$\eta\vert_{I_{LOAD,max}}$	[%]	99.998	99.993	99.918	99.986	99.979	99.926	99.999
Area	[mm^2]	0.09[b,c]	0.064	0.017	0.038	0.21	0.018	0.195
Response time[a]	[μs]	4.0	0.15	6.0	2.7	0.65	0.4	731.6
ΔV_{OUT} varying V_{IN}								
• Maximum	[mV]	200[d]	196	8.91	–[e]	–[e]	–[e]	246
• Minimum	[mV]	−120[d]	−183	−10.63	–[e]	–[e]	–[e]	−281
$\Delta V_{IN}/t_r$[f]	[V/μs]	0.5/1[d]	0.5/0.5	0.2/10	–[e]	–[e]	–[e]	0.6/1
ΔV_{OUT} varying I_{LOAD}								
• Maximum	[mV]	200[d]	231	0.00	200[d]	200	56	274
• Minimum	[mV]	−425[d]	−243	−68.8	−270	−385	−42	−368
$\Delta I_{LOAD}/t_r$[f]	[mA/μs]	49.50/0.2[d]	99.95/0.5	100/0.3	99.9/1	199.5/0.5	49.95/0.2	99.9/1
Line regulation	[mV/V]	3.625	1.0	4.70	–[e]	8.9	8.1	2.50
Load regulation	[μV/mA]	148.0	80.0	300.0	–[e]	108.0	55.6	85.44

[a]Worst case, [b]Effective area, [c]Pads included, [d]Estimation based on published results, [e]Not available, [f]t_r: Rise time

Table 4.6 FOM_1 values obtained by the regulators considered in Table 4.5

		[47]	[53]	[54]	[55]	[56]	[80]	This work
T_r	[ns]	1.25	0.47	0.07	0.47	0.12	0.04	0.64
FOM_1	[fs]	30.00	33.18	56.69	65.80	24.28	29.01	3.76

less devices. This chapter presents the design of an ultra-low-power IC-LDO regulator.

The proposed solution uses the classical topology of LDO regulators presented in Chap. 2, where a class AB voltage buffer has been included between the error amplifier and the pass transistor to provide the large currents required to enhance the transient response while maintaining ultra-low quiescent power consumption.

Using a class AB buffer relaxes the requirements of the error amplifier, making it possible to build a low-power topology which complies with the gain and bandwidth constraints, and provides good line and load regulations.

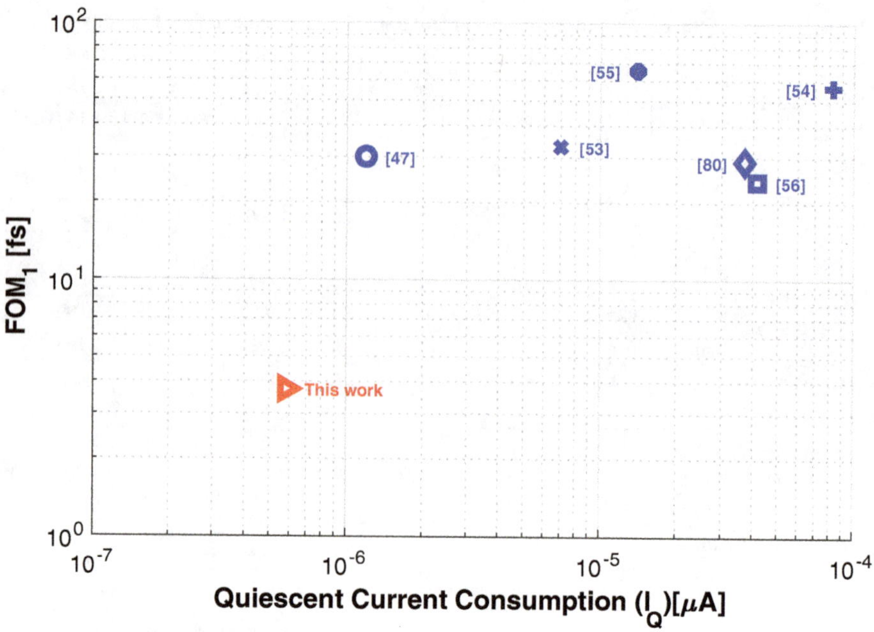

Fig. 4.13 FOM_1 of the regulators in Table 4.6 versus quiescent current consumption

Chapter 5
The Flipped Voltage Follower (FVF): An Alternative Topology for LDO Regulators

Abstract To overcome the limitations of the classical topology to build IC-LDO regulators, several authors have chosen alternative topologies. This is the case of the Flipped Voltage Follower cell (proposed by Ramirez-Angulo et al., ISCAS 2002), which meets the requirements of LDO regulators as a consequence of its low output impedance and good stability. The first use of this cell as part of a linear regulator due to Pulkin and Rincon-Mora (U.S. Patent No. 6,573,694, 2003), where the Cascode Flipped Voltage Follower cell was used as a buffer to drive the pass transistor. Later, this same cell was used in an LDO regulator as a power stage (Hazucha et al., JSSC 40(4), 2005). This chapter describes the FVF and CAFVF cells and their performances as an LDO regulator, highlighting their advantages and disadvantages when compared to the classical topology. Then, it offers a thorough review of the regulators presented in the literature that use the FVF family of cells. Finally, a new regulator is presented that improves the performances of CAFVF-based regulators both in regulation and transient response.

5.1 Introduction

The FVF cell, Fig. 5.1a, is an evolution of the classical common-drain amplifier which is also known as Voltage Follower. Although it had been already used in some circuits, it was not until [66, 67] when this cell was identified and coined as the Flipped Voltage Follower. Its main advantage comes from the fact that, due to local feedback, it performs a very low output impedance, $R_{OUT} = 1/g_{m,EA}g_{m,P}r_{o,EA}$. However, despite the FVF-based LDO regulator is able to supply a high output current, its maximum sinking current is limited by the biasing current $I_{BIAS,1}$, which directly affects the transient response of the regulator.

Figure 5.2 shows a comparison between the classical topology of an LDO regulator and the FVF-based solution. In Fig. 5.2b, M_{PASS} is the pass transistor that provides the required current to the load. M_{EA} is a common-gate amplifier which acts as an error amplifier and compares the output voltage, V_{OUT}, to a reference voltage, V_{REF}, generating the control voltage, V_{GATE}, at the gate of M_{PASS}.

© Springer International Publishing AG, part of Springer Nature 2018

J. M. Hinojo et al., *Internally Compensated LDO Regulators for Modern System-on-Chip Design*, https://doi.org/10.1007/978-3-319-75411-6_5

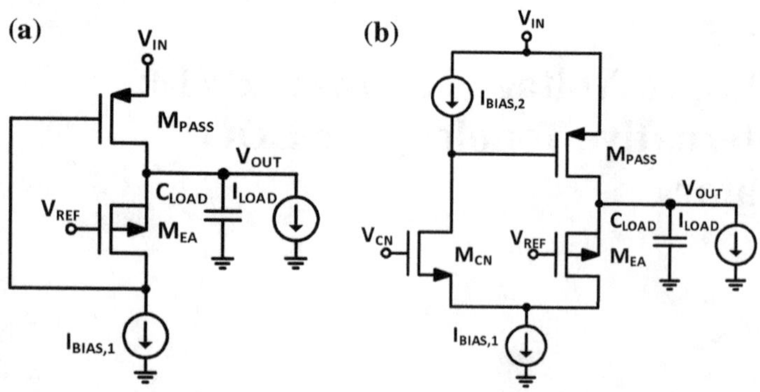

Fig. 5.1 Schematics of IC-LDO regulators based on: **a** FVF and **b** CAFVF cells

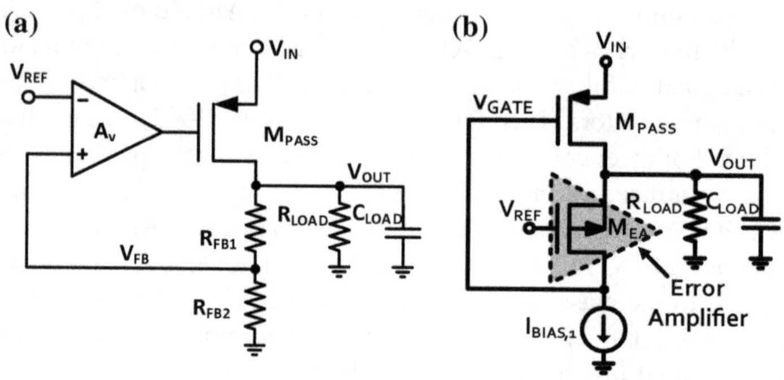

Fig. 5.2 Comparison between: **a** classic topology and **b** FVF-based IC-LDO regulator

The regulation mechanism provided by the local feedback is graphically depicted in Fig. 5.3. Starting from steady-state conditions, if I_{LOAD} increases, V_{OUT} will decrease, and, due to the positive gain of the common-gate transistor M_{EA}, V_{GATE} will also decrease. As a result, the current delivered to the load will increase, which compensates for the changes produced by the initial perturbation. On the other hand, if I_{LOAD} increases, V_{OUT} will also increase producing an increase of V_{GATE} and a decrease in the output current.

V_{REF} is related to V_{OUT} through Eq. 5.1 where $V_{SG,EA}$ is the source-to-gate voltage of the transistor acting as an error amplifier. In a first-order approach, this voltage can be considered to be constant and independent on I_{LOAD}.

$$V_{OUT} = V_{REF} + V_{SG,EA} \tag{5.1}$$

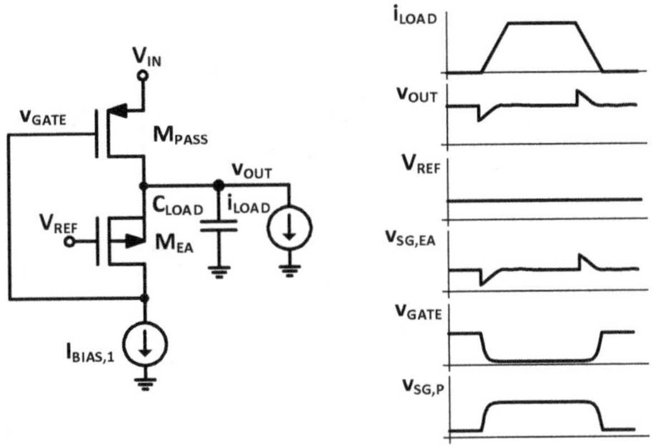

Fig. 5.3 Regulation mechanism of the FVF-based LDO

Note that the transient response of this regulator is limited by the value of $I_{BIAS,1}$, which is the maximum current that the regulator can sink from the load. In addition, the low gain of the FVF cell limits the line and load regulation.

5.2 FVF- and CAFVF-Based LDO Regulators

References [155, 156] use, for the first time, the FVF as the LDO regulator core. Authors make use of the basic structure shown in Fig. 5.1a to generate the regulated output voltage V_{OUT}. A unit gain amplifier is used to generate V_{REF} (Fig. 5.4), where V_{BG} is a voltage reference connected to the input of the unit gain amplifier and V_{REF} is given by Eq. 5.2.

$$V_{REF} = V_{BG} - V_{SG,EA1} \tag{5.2}$$

As the size of the diode-connected transistor $M_{EA,1}$ is equal to M_{EA}, and both are biased with the same bias current $I_{BIAS,1}$, then $V_{SG,EA1} = V_{SG,EA}$ and $V_{OUT} = V_{BG}$. The circuit on the right of Fig. 5.4 is a direct implementation of the FVF cell of Fig. 5.1a. Therefore, it has the same limitations regarding transient response, line and load regulations.

Later, in references [110, 158, 159], the authors replace the pass transistor in the FVF by the so-called composite transistor [157]. This cell, shown in Fig. 5.5, is a large bandwidth amplification stage [160], which controls M_{PASS}. The additional gain stage is intended to drive the parasitic capacitance at the M_{PASS} gate and increase the open-loop gain. Moreover, the equivalent impedance at the gate of the pass transistor is approximately $R_{eq\_cgs}|^{M_{PASS}} = 1/g_{m,M2}$ [157]. This allows to move the location of the pole formed by $R_{eq\_cgs}|^{M_{PASS}}$ and C_{GATE} to higher frequencies, enhancing the stability of the system and increasing its bandwidth. In Fig. 5.6, the structure of the resulting IC-LDO regulator is shown. In this regulator, and with the aim of improving

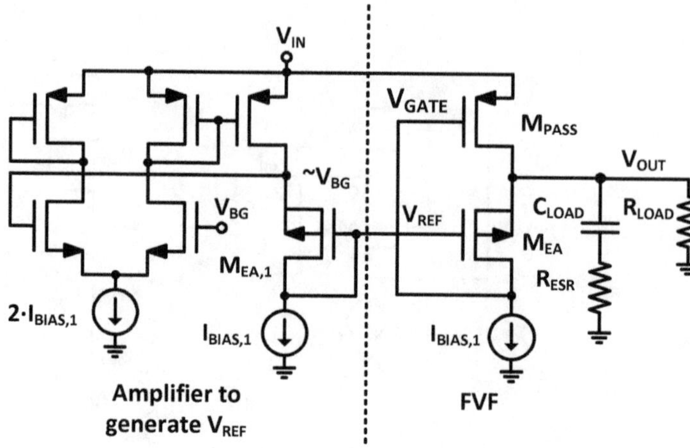

Fig. 5.4 Schematic of the LDO regulator proposed in [156]

Fig. 5.5 The so-called composite transistor implementation [157]

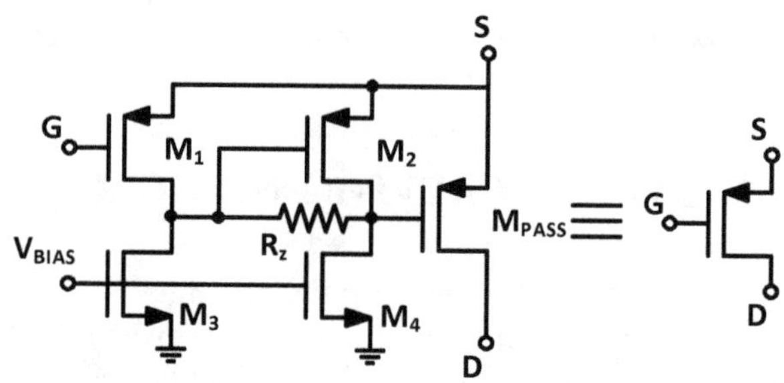

its stability against changes in C_{LOAD}, reference [110] adds an additional branch that makes it able to drive load capacitors up to 100 nF.

Another author proposes in [161] the use of a cascode transistor to improve the open-loop gain and, consequently, the load regulation (Fig. 5.7). The gain of the common-gate amplifier formed by M_{EA} and M_{N1} is

$$A_{CG} = 1 + (g_{m,EA} + g_{m,N1})r_{0,EA} \| r_{0,N1} \tag{5.3}$$

where $g_{m,EA}$ is the transconductance of M_{EA}, and $r_{o,EA}$ and $r_{o,N1}$ are the output resistances of transistors M_{EA} and M_{N1}, respectively. Thus, the total resistance at the output of the LDO regulator is

$$R_{OUT} = \frac{R_{LOAD} \| r_{o,P}}{1 + g_{m,P}(R_{LOAD} \| r_{o,P}) + g_{m,P}(R_{LOAD} \| r_{o,P})(g_{m,EA} + g_{m,N1})(r_{o,EA} \| r_{0,N1})} \tag{5.4}$$

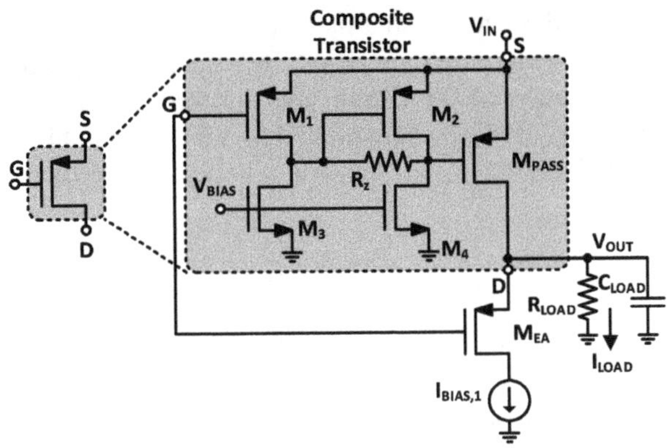

Fig. 5.6 Structure of the LDO regulator proposed in [158] based on the FVF cell that makes use of the composite transistor

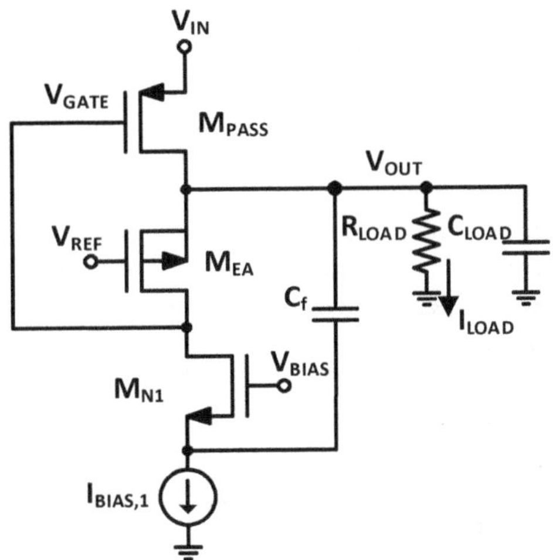

Fig. 5.7 Structure of the LDO regulator proposed in [161] based on the FVF cell

Hence, the open-loop gain is

$$A_{OL} = g_{m,P} R_{OUT} g_{m,EA} \left(r_{o,EA} \| r_{o,N1} \right) = \\ = \frac{g_{m,p} g_{m,EA} (R_{LOAD} \| r_{o,P})(r_{o,EA} \| r_{o,N1})}{1 + g_{m,p}(R_{LOAD} \| r_{o,P}) + (g_{m,p}(R_{LOAD} \| r_{o,P})(g_{m,EA} + g_{m,N1})(r_{o,EA} \| r_{o,N1})} \quad (5.5)$$

Despite the benefits of including transistor M_{N1}, it increases the minimum input voltage in one $V_{DS,sat}$, resulting in $V_{IN,min} = V_{SG,p} + 2V_{DS,sat}$, which is not suitable for low voltage designs or nanometric technologies.

A way to improve the open-loop gain of the FVF cell, without limiting the input voltage range is the use of its cascoded version, the so-called CAFVF cell depicted in Fig. 5.1b. Furthermore, the main advantage of this cell comes from the increase of the total open-loop gain due to the inclusion of transistor M_{CN} in the local feedback loop [154].

According to its small-signal model, this new transistor is seen as a second common-gate amplifier. Thus, its effect is added to M_{EA}. Another advantage is the reduction of the output impedance, Eq. 5.6, which has been obtained assuming $r_{B1}, r_{B2} \gg r_{o,x}$, where r_{B1} and r_{B2} are the output impedances of $I_{BIAS,1}$ and $I_{BIAS,2}$, respectively, and $r_{o,x}$ is the output impedance of transistor M_x.

$$R_{OUT} \approx \frac{1}{g_{m,P} g_{m,EA} g_{m,CN} r_{o,EA} r_{o,CN}} \tag{5.6}$$

One of the first references that use the CAFVF cell as an internally compensated LDO regulator is [87]. Therein M_{PASS} is replaced by a group of CAFVF cells. Figure 5.8 shows the block diagram of this solution, as well as the implementation of each CAFVF cells, which replaces M_{PASS}. Each stage was designed to drive a maximum current of 5 mA, so that 20 stages were required to provide a maximum output current of 100 mA.

Since the output voltage is not observed by the feedback loop, Fig. 5.8a, to achieve a fast load regulation, a replica output stage is included in this loop to regulate V_{OUT}. By means of the error amplifier A_v, and a buffer that drives the CAFVF cells, a low-noise control voltage V_{CTRL} is generated in order to ensure that V_{OUT} matches the reference voltage for $I_{LOAD} = 0$ A. For non-zero load current, the CAFVF cell provides the required regulation mechanism in order to keep the output voltage spikes, peak-to-peak, under a 10% of the nominal value of V_{OUT}, which is imposed by the application.

The main drawback of this structure is the need of a large capacitor, 600 pF, to stabilize the system and filter out the external noise. This capacitor occupies approximately 90% of the total area of 0.1 mm^2 in a 90-nm CMOS technology design. Therefore, and despite the good transient response of this regulator, it is not suitable SoCs.

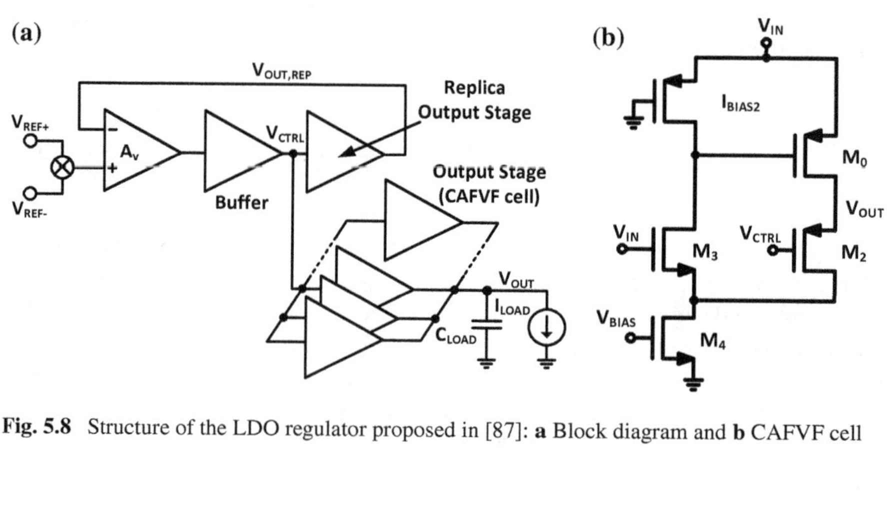

Fig. 5.8 Structure of the LDO regulator proposed in [87]: **a** Block diagram and **b** CAFVF cell

As in the case of the FVF, the transient response of the CAFVF is limited by $I_{BIAS,1}$ and $I_{BIAS,2}$. Therefore, a high value for these bias currents is required to provide a good transient response, which leads to high quiescent power consumption and poor efficiency. In [162], authors use RC couplings to dynamically increase the corresponding bias currents when the output voltage changes (Fig. 5.9). It reduces the charge/discharge time of C_{GATE}. As a consequence, the transient response improves without detriment to the quiescent power consumption.

A similar solution is proposed in [163], where the authors add two amplifiers to the gate of transistors $M_{BIAS,1}$ and $M_{BIAS,2}$, respectively, to amplify the effect of the RC couplings, with the aim of boosting the corresponding current ($I_{BIAS,1}$ or $I_{BIAS,2}$), enhancing the slew-rate of the CAFVF cell.

In [164], a new mechanism to charge/discharge the parasitic capacitance of M_{PASS} is proposed, Fig. 5.10. A digitally controlled push–pull stage is connected to the pass transistor gate (node A). This stage detects changes at nodes A and B due to I_{LOAD} variations. The detection of these events is based on the comparison of the voltage signal at node A (B) with a delayed version of itself. In this way, a pulse is generated that activates the corresponding transistor of the push–pull stage, which is composed of transistors M_{DN} and M_{UP}, during the transient response, deactivating it when the nominal condition is attained. The inverter chain is designed to delay the input signal a time long enough to ensure the stability of the regulator, avoiding undesired switching.

The charging/discharging of C_{GATE} can be improved if a buffer is introduced. Figure 5.11a, b depict the resulting circuit for the FVF and the CAFVF, respectively. Unfortunately, when an additional element is introduced in the signal path, the order of the system increases and, consequently, the stability deteriorates.

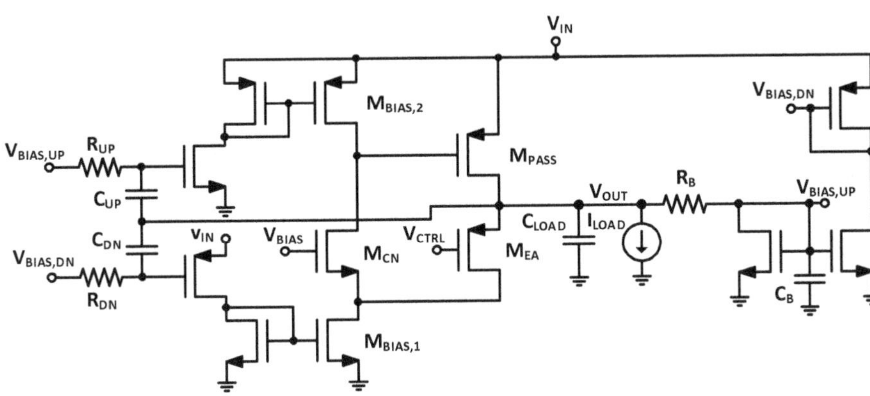

Fig. 5.9 Structure of the CAFVF-based LDO regulator proposed in [162] including capacitive couplings to improve the transient response

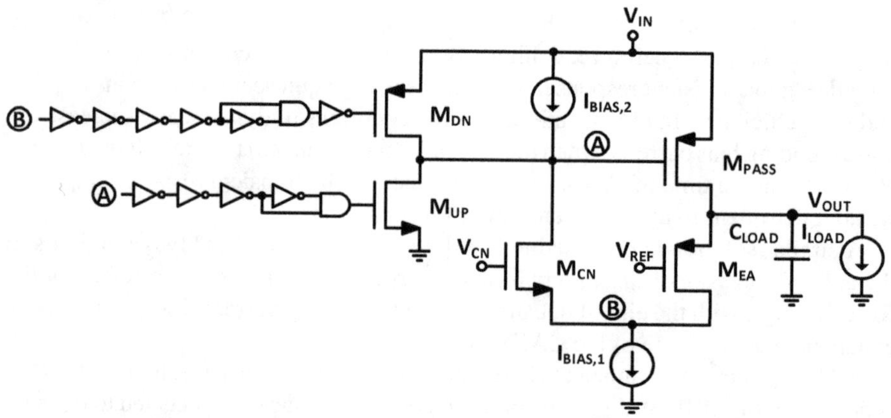

Fig. 5.10 Structure of the IC-LDO regulator proposed in [164] that uses digital spike detection to control a push–pull stage in order to improve the transient response

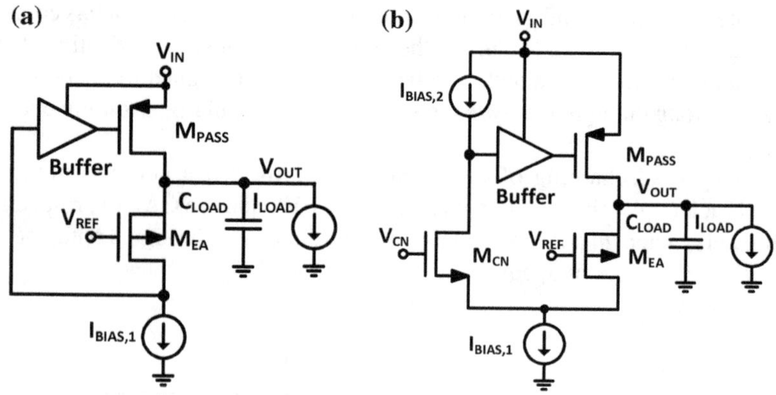

Fig. 5.11 Inserting a buffer to drive the parasitic capacitance at the gate of M_{PASS} in an IC-LDO regulator based on: **a** the FVF and **b** the CAFVF cells

An example of this approach is found in [88, 165], where the authors use a common source non-inverting amplifier as a buffer in the CAFVF cell. The amplifier allows to increase the open-loop gain and the voltage swing at the gate of M_{PASS}. Moreover, to improve the slew-rate at the gate of M_{PASS}, capacitive coupling is used to detect variations of V_{OUT} and to dynamically modify the current at the buffer input.

Authors in [166, 167] propose a circuit that can be used as an LDO regulator and as a digital power gate [168], providing an output current up to 4 A. Figure 5.12 depicts

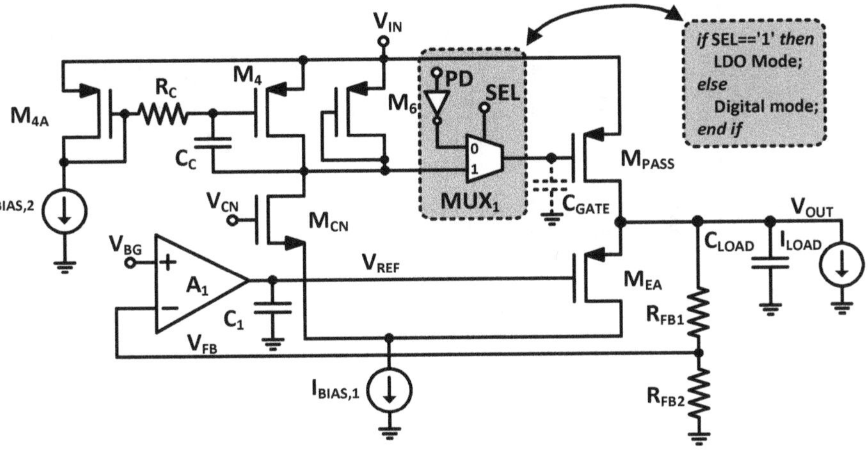

Fig. 5.12 Structure of the LDO regulator proposed in [166]

the block diagram of the proposed regulator which includes a multiplexer, MUX_1, to select the operation mode. In this way, in LDO mode, the circuit implements two feedback loops. One, formed by the operational amplifier A_1 and C_1, is in charge of fixing the value of the desired V_{OUT}. The other, formed by the CAFVF cell, will react to compensate for fast variations of V_{OUT}.

From the stability point of view, authors implement a compensation scheme consisting of transistors M_4, M_6 and the passive components R_C and C_C. The rationale behind this scheme is to generate a phase delay network that will compensate for the change in the location of the non-dominant pole due to variations in C_{LOAD}. Note that M_4, R_C and C_C make a capacitor multiplier at low frequencies, and thus, it allows reducing the required value of C_C, usually chosen as several times larger than C_{GATE}. Moreover, M_6 is included to limit the total open-loop gain of the regulator to ensure its stability.

Table 5.1 compares the main performances of the solutions previously discussed in this chapter. With the aim of a fair comparison, the FOM_1, defined in Eq. 2.24, has been used and the values obtained for each regulator are also included in this table. The best FOM_1 is achieved by the work proposed in [88], due to its low quiescent power consumption and low C_{LOAD}. Figure 5.13 depicts the FOM_1 versus quiescent current consumption for a large number of LDO regulators based on the FVF and derived cells, and those regulators based on the classical LDO topology (Chap. 2).

Table 5.1 Comparison of IC-LDO regulators based on the FVF and derived cells

		[87]	[88]	[110]	[156]	[159][a]	[161]	[163]	[164]	[166]
Process	$[\mu m]$	0.09	0.09	0.065	0.35	0.065	0.35	0.35	0.18	0.014
V_{IN}	[V]	1.2	0.75–1.2	1.2	1.2–1.5	1.2	1.4–3.3	1.28–3.3	0.9–1.8	0.5–1.1
V_{OUT}	[V]	0.9	0.5–1	1.0	1.0	1.0	1.2	1.1	0.7	0.4–1.0
$V_{DROPOUT}$	[mV]	300	200	200	200	200	200	180	200	80
$I_{LOAD,max}$	[mA]	100	100	50	50	50	50	100	50	4000
I_Q^{b}	[μA]	69000	8	23.7	95	13.2	34.69	25	3.9	10000
C_{OUT}	[pF]	6e2	5e1	1e4	1e2	1e4	2e2	1e2	1e2	7e6
$\eta\|I_{LOAD,max}$	[%]	94.340	99.992	99.953	99.905	99.987	99.965	99.975	99.996	99.751
Area	$[mm^2]$	0.098[c]	0.019	–[c]	0.045	–[d]	0.080	0.126	0.041	0.041[f]
Response time[b]	$[\mu s]$	0.015[c]	3.75[c]	1.65	0.3	0.925	1.4[c]	1.4	–[d]	1.5[c]
ΔV_{OUT} varying V_{IN}										
• Maximum	$[mV]$	–[d]	40	–[d]	–[d]	–[d]	20[c]	20[c]	–[d]	–[d]
• Minimum	$[mV]$	–[d]	−33	–[d]	–[d]	–[d]	−28[c]	0[c]	–[d]	–[d]
$\Delta V_{IN}/t_r^{e}$	$[V/\mu s]$	–[d]	0.42/10	–[d]	–[d]	–[d]	0.2/0.3	1/1e3	–[d]	–[d]
ΔV_{OUT} varying I_{LOAD}										
• Maximum	$[mV]$	45	114	19	130	43.92	46	31	500[c]	35[c]
• Minimum	$[mV]$	−45	−73	−58	−160[c]	−341.63	−75	−80	−500[c]	−40[c]
$\Delta I_{LOAD}/t_r^{e}$	$[mA/\mu s]$	–[d]	98.5/0.1	50/0.1	50/0.5	50/0.1[c]	49.9/0.3	100/0.5	50/0.1	3650/1.5
Line regulation	$[mV/V]$	–[d]	3.78	8.89	18	216.8	8.8	–[d]	–[d]	–[d]
Load regulation	$[\mu V/mA]$	900[a]	100	34	280	133e3	3000	190[c]	–[d]	–[d]
T_r	[ns]	0.54	0.09	15.4	0.58	77.11	0.48	0.11	2.0	131.25
FOM_1	[fs]	32400.00	7.48	7299.6	1102	20357.04	334.93	27.75	156.00	328125.00

[a]Simulation results, [b]Worst case, [c]Estimation based on published results, [d]Not available, [e]t_r: Rise time, [f]Pass transistor not included

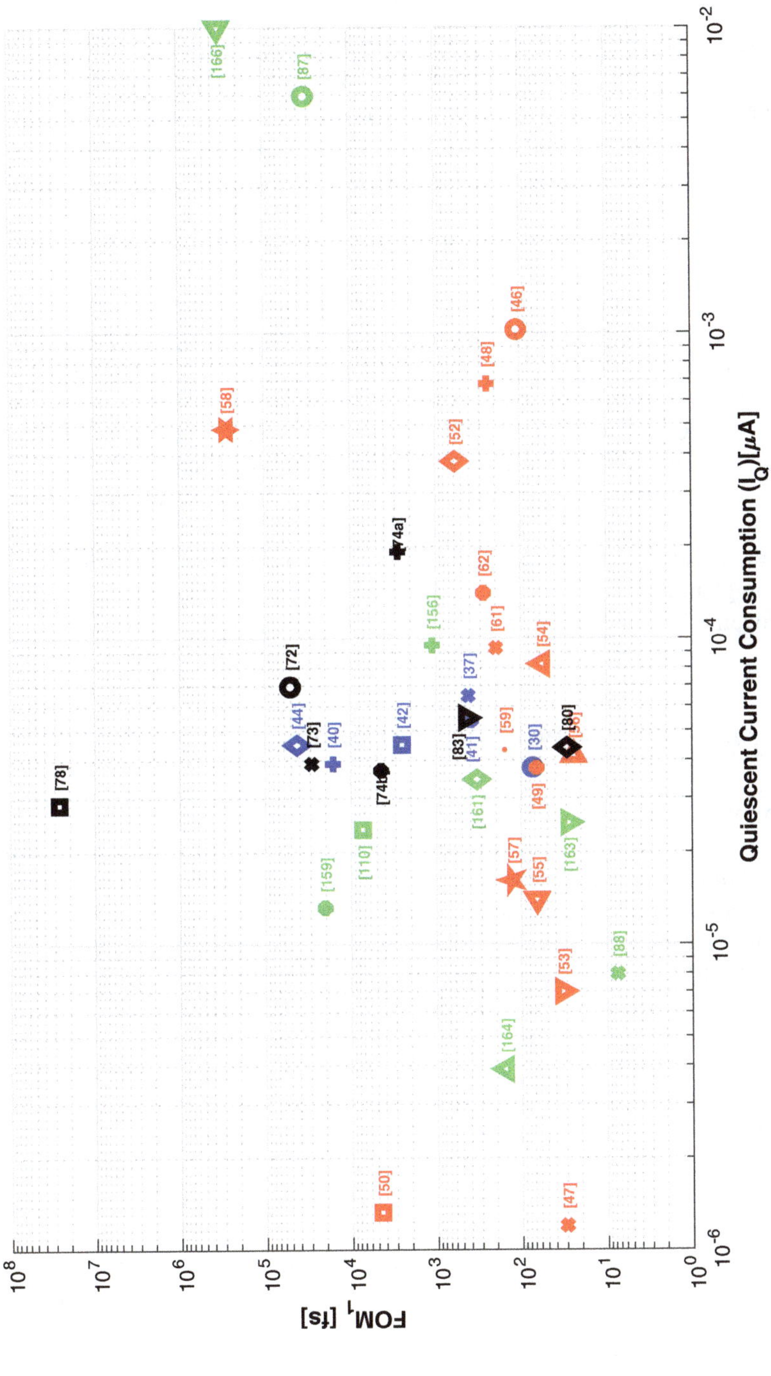

Fig. 5.13 FOM_1 versus quiescent current consumption

5.3 Design of an FVF-Based IC-LDO Regulator

As it was discussed in previous sections, the transient response of an internally compensated LDO regulator based on the FVF or CAFVF cells is limited by the value of the static biasing current sources, as they set the time required to charge/discharge C_{GATE}. Increasing the regulator speed implies increasing the value of these current sources, and consequently, the total quiescent power consumption, which directly impacts on the regulator power efficiency. In this section, a new structure based on capacitive coupling is proposed to implement dynamic biasing in an LDO regulator based on the CAFVF cell.

To improve the transient response of an LDO regulator capacitive coupling has been successfully used in, for instance, [161–164, 167] or [166]. These references make use of one or several capacitive couplings to improve the transient response to load variations. However, none of them addresses the degradation in the transient response due to line variations. This section describes the design of an IC-LDO regulator in a standard 65-nm CMOS technology with the same specifications of the regulator designed in Chap. 3. It uses the CAFVF cell with capacitive couplings combined with a transistor biased in the edge from saturation to ohmic region to improve the transient response, and a boosting stage that increases the regulator open-loop gain. It will be shown that the combination of these techniques achieves an improvement in line and load regulations, as well as in the transient response.

5.3.1 Structure and Principle of Operation

The core of the proposed LDO regulator is the CAFVF cell shown in Fig. 5.1b. In order to improve the poor settling time of the CAFVF under low-power constraints, this contribution proposes an alternative version of this cell that enhances the transient response for load and line variations without negative effects on the quiescent power consumption or the stability of the circuit. Figure 5.14 depicts the complete scheme of the proposed LDO regulator, where the gain of the regulation loop is increased by means of the gain-boosting amplifier A_0 [145] without degrading the circuit speed. To avoid instability, this amplifier is designed according to the method explained in [169]. A compensation capacitor for A_0 is not required because the parasitic gate-source capacitance of M_{CN} is large enough, and approximately constant.

In addition, transient line and load responses are improved by dynamically increasing the currents responsible for charging/discharging C_{GATE}. In particular, $I_{BIAS,1}$ is replaced by a charge-fast settling path (C-FSP) that is formed by a dynamic current source that increases the charging current when the input or output voltage increases. This block is implemented by transistors M_7–M_{14} and the gain-boosting amplifier A_1. For A_1, the design considerations aresimilar to those used for A_0. When the voltage V_{IN} rises, V_{OUT} instantaneously tends to grow. Thus, V_{GATE} must increase rapidly to recover the nominal value of the output voltage.

To this end, a high transient current $I_{BIAS,1}$ is provided by means of RC coupling, $R_1 - C_1$, which increases the positive Slew-Rate (SR+), at the gate of M_{PASS}. In addition, the output voltage is also coupled through C_2 to magnify this effect, because V_{IN} and V_{OUT} tend to exhibit a similar behaviour.

A symmetrical discharge-fast settling path (D-FSP) was included to dynamically increase the discharging current of C_{GATE} when the input or output voltages decrease. Note that, in this case, an additional inverting amplifier, A_2, is required. This is implemented with a simple, low power, differential pair.

In order to reduce the static power consumption, A_2 is biased in the sub-threshold region and M_{35}-M_{38} are biased in the edge from saturation to the ohmic region. When V_{IN} or V_{OUT} decreases, the current through transistor M_{38} rapidly increases owing to a change in its operating region from saturation to the ohmic region, as described in [170]. The multiplying factor $K = 1 : 5.5$ of the current mirror composed of M_{37}-M_{38} (W/L = 20 μm/0.12 μm) and M_{39}-M_{40} (W/L = 110 μm/0.12 μm), generates a high transient current in the V_{GATE} branch, and consequently produces a large negative Slew-Rate (SR−) at the gate of M_{PASS}. Figures 5.15 and 5.16 depict the large transient charging and discharging currents that enhance the slew-rate. In these figures, $i_{AB} = I_{BIAS,2} - (i_B + i_{CM1}) \approx I_{BIAS,2} - i_B$, where i_B, is the total current that flows through M_{CN} and M_{40}.

Finally, in order to reduce the overshoot when the load current is switched from heavy to light load, capacitor C_3, resistor R_2 and transistor M_4 are added. Note that, under steady-state conditions, M_4 is in the cut-off region, but when I_{LOAD} decreases, R_2 and C_3 sense the voltage spikes from V_A (Fig. 5.14) and couple them to the gate of M_4. This momentarily increases the discharging current. As a consequence, the magnitude of the overshoot is reduced. Transistor M_4 is sized to sink enough current to maintain the overshoot under 10% of the nominal output voltage without a significant increment in the total area of the proposed LDO regulator.

Note that the resistor values, R_1–R_3, are chosen to move the RC coupling effects towards a high frequency. For these values, capacitors C_1–C_5 are calculated to achieve the appropriate increment of V_{GATE}, Eq. 5.7, in order to provide the required current. Table 5.2 lists the values selected for these components.

$$C = -\frac{R}{\Delta t} ln\left(1 - \frac{\Delta V_{GATE} \Delta t}{\Delta V_{IN}}\right) \tag{5.7}$$

Biasing voltages are generated by the circuit shown in Fig. 5.17. Each branch is formed by transistors in a single-diode connection, $M_{P1,VCP}$ and $M_{P2,VCP}$ or $M_{N1,VCN}$ and $M_{N2,VCN}$ for PMOS and NMOS versions, respectively, and by a cascode current source, M_{BP1} or M_{BN1}. Specifically, the aspect ratios of $M_{P1,VCP}$ or $M_{N1,VCN}$, which operate in the triode region, are chosen to be lower than those of $M_{P2,VCP}$ or $M_{N2,VCN}$, in order to create the required gate voltage to supply the cascode transistor. V_{BIAS} in Fig. 5.14 is an external source. The current consumption of the biasing circuitry

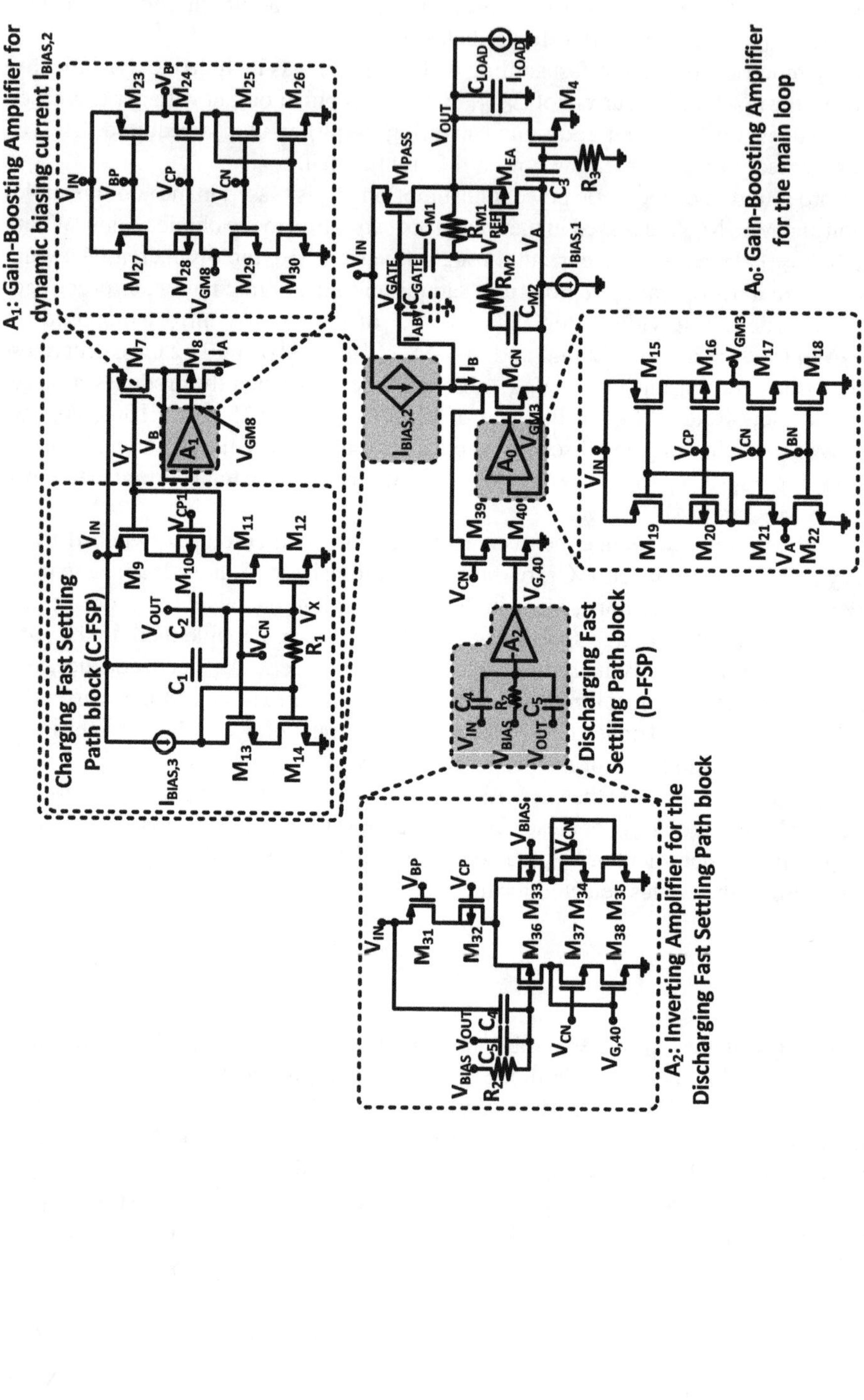

Fig. 5.14 Structure of proposed IC-LDO regulator: Circuit core showing those blocks that improve settling time and limit overshoot. A detailed view of the implementation of amplifiers A_0, A_1 and A_2 is also included

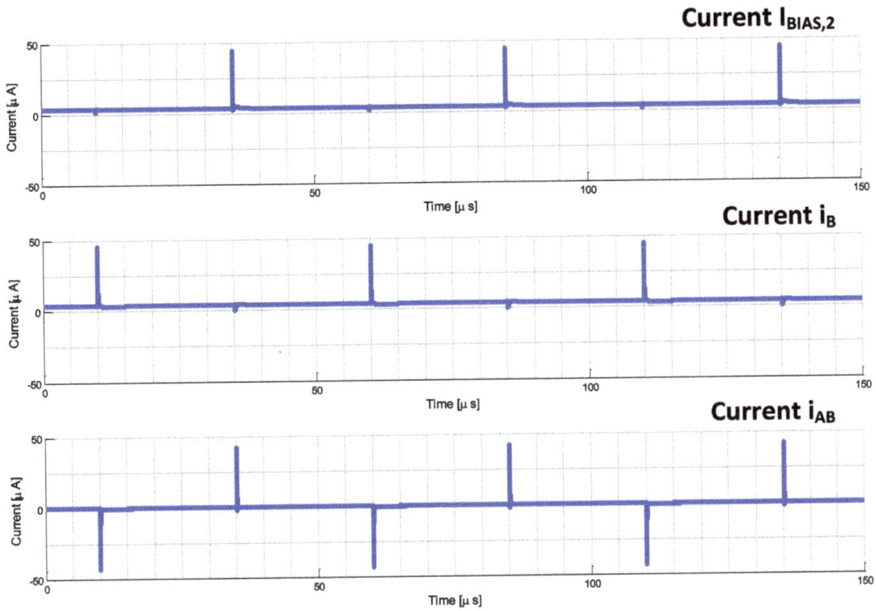

Fig. 5.15 Dynamic behaviour of $I_{BIAS,2}$, I_B and I_{AB} (Fig. 5.14) when I_{LOAD} changes following a square wave between its minimum (0.1 mA) and maximum (100 mA) values, with rise and fall times of 1 μs

is 3.6 μA, as $I_{BIASING}$ is chosen to be 100 nA. Table 5.3 lists the transistor sizes and current ratios for the circuit in Fig. 5.17.

5.3.2 *Stability Analysis*

In this section, the stability analysis of the proposed LDO regulator is studied. The major concern regarding stability for IC-LDO regulators is raised by the large load current variations. These variations produce significant changes in the small-signal parameters, which affect the location of poles and zeros. Fortunately, this is not the case for line voltage variations. A simplified small-signal model of the proposed structure is depicted in Fig. 5.18, where, as usual, $g_{m,i}$ and $r_{o,i}$ are the transconductance and output resistance of transistor M_i, respectively. In addition, as it was defined in Chap. 4, $R_{GB,NMOS}$ and $R_{GB,PMOS}$ represent the equivalent output resistance of amplifiers A_0 and A_1, respectively. $C_{gg,NMOS}$ and $C_{gg,PMOS}$ model the parasitic capacitance at the gate of M_8 and M_{CN}. The poles and zeros derived from the RC couplings are neglected because they are located at a high frequency.

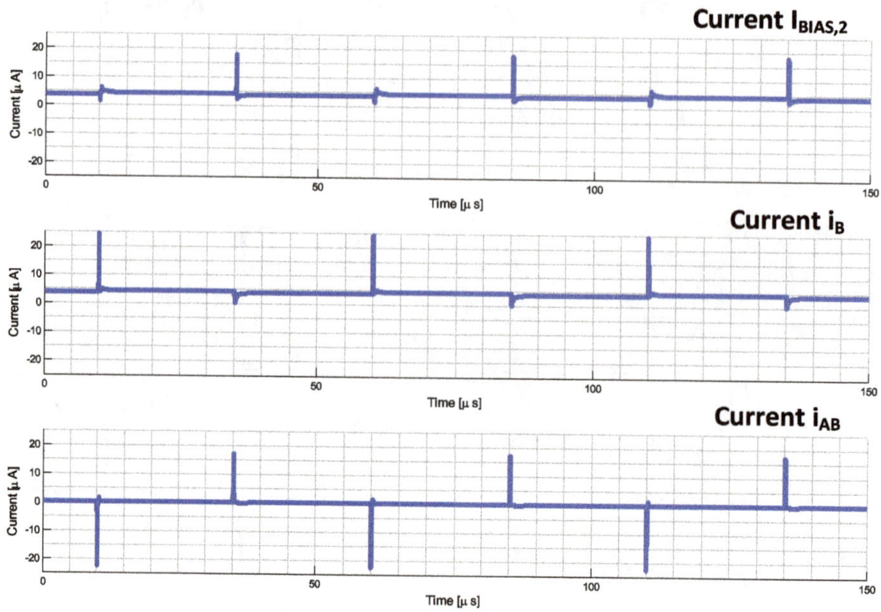

Fig. 5.16 Dynamic behaviour of $I_{BIAS,2}$, I_B and I_{AB} (Fig. 5.14) when V_{IN} changes in a square wave between its minimum (0.9 V) and maximum (1.2 V) values, with rise and fall times of 1 μs

Table 5.2 Selected values used for RC couplings

Device	Value	Device	Value [pF]
R_1	100 [kΩ]	C_2	1.25
R_2	335 [kΩ]	C_3	5.0
R_3	100 [kΩ]	C_4	0.125
C_1	0.125 [pF]	C_5	1.25

Table 5.3 Multiplying factors and aspect ratios for transistors in biasing circuit

Transistor	Aspect ratio (Width [μm]/Length [μm])	Current ratio	Value
$M_{N1,CN}$	0.14/1.39	M_0	4
$M_{P1,VCP}$	0.12/1.49	N_0	4
M_{BN1}–M_{BN2}	0.14/0.12	–	–
M_{BP1}–M_{BP2}	0.30/0.12	–	–

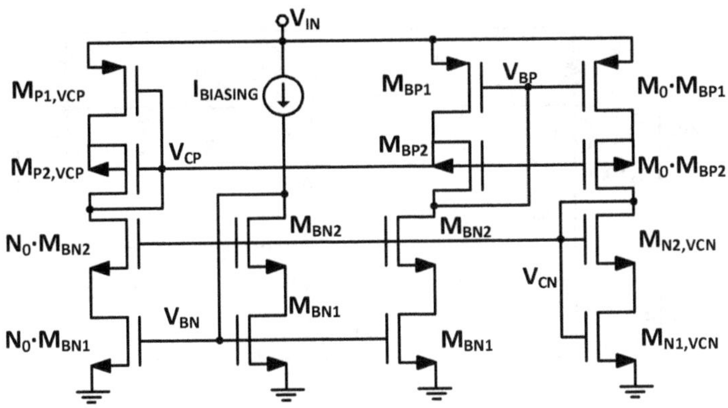

Fig. 5.17 Cascode voltages and V_{BIAS} biasing circuits

Concerning the transfer function in Eq. 5.8, the DC gain is approximated by Eq. 5.9. Approximate values for the transfer function coefficients are given in Eqs. 5.10–5.22.

$$H(s) = A_{OL}\frac{1 + a_1 s + a_2 s^2 + a_3 s^3 + a_4 s^4 + a_5 s^5 + a_6 s^6}{1 + b_1 s + b_2 s^2 + b_3 s^3 + b_4 s^4 + b_5 s^5 + b_6 s^6 + b_7 s^7} \tag{5.8}$$

where

$$A_{OL} = g_{m,EA} g_{m,P} R_{out,D-FSP}(R_{LOAD} \| r_{o,P}) \tag{5.9}$$

$$a_1 = R_3 C_3 \tag{5.10}$$

$$a_2 = R_3 (R_{M2} + R_{M1}) C_3 C_{M2} \tag{5.11}$$

$$a_3 = R_3 R_{M2} C_3 C_{M2} \left(R_{M1} C_{M1} + \frac{R_{GB,NMOS} C_{gg,NMOS}}{A_0} + \frac{R_{GB,PMOS} C_{gg,PMOS}}{A_1} \right) \tag{5.12}$$

$$a_4 = \frac{1}{A_0} R_3 R_{M1} R_{GB,NMOS} C_3 C_{M2} C_{M1} C_{gg,NMOS} \left(\frac{1}{g_{m,CN}} + R_{M2} \right) \tag{5.13}$$

$$a_5 = \frac{1}{A_1 A_0} R_3 R_{M1} R_{GB,NMOS} R_{GB,PMOS} C_3 C_{M2} C_{M1} C_{gg,NMOS} C_{gg,PMOS} \left[\frac{1}{g_{m,CN}} + R_{M2}\left(1 + \frac{1}{g_{m,CN} r_{o,CN}}\right) + \frac{1}{g_{m,8} g_{m,CN} r_{o,7}} \right] \tag{5.14}$$

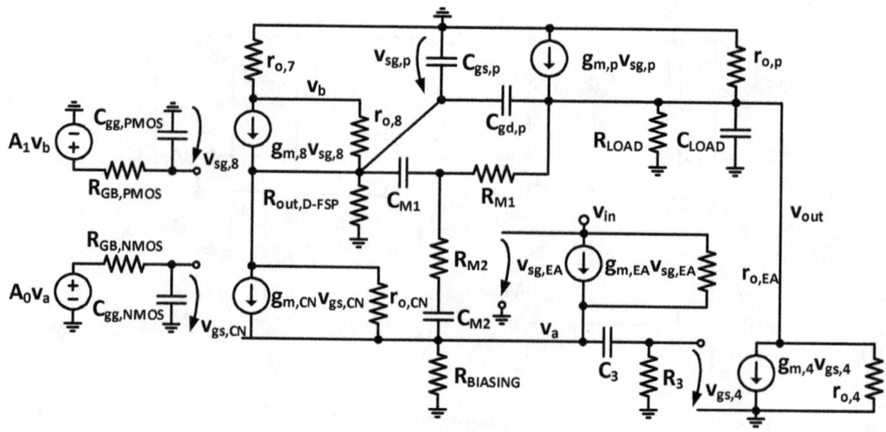

Fig. 5.18 Small-signal model of the proposed IC-LDO regulator

$$\begin{aligned} a_6 = {} & \frac{1}{g_{m,P}A_1A_0}R_3R_{M1}R_{GB,NMOS}R_{GB,PMOS}C_3C_{M2}C_{M1}C_{gd,P}C_{gg,NMOS}C_{gg,PMOS}\left[R_{M2}+\right. \\ & \left. + \frac{1}{g_{m,CN}} + \frac{1}{g_{m,8}g_{m,CN}}\left(\frac{1}{r_{o,8}}+\frac{1}{r_{o,7}}\right)\right] \end{aligned} \tag{5.15}$$

$$b_1 = g_{m,P}R_{out,D-FSP}R_{LOAD}\|r_{o,P}\left(C_{M2}+C_{M1}+C_{gd,P}\right) \tag{5.16}$$

$$b_2 = g_{m,P}R_{out,D-FSP}R_3R_{LOAD}\|r_{o,P}\left(C_{M2}+C_{M1}+C_{gd,P}\right)C_3 + R_3R_{out,D-FSP}C_3C_{M1} \tag{5.17}$$

$$\begin{aligned} b_3 = {} & g_{m,P}R_{out,D-FSP}R_3R_{LOAD}\|r_{o,P}C_{M2}C_3\left[R_{M2}\left(C_{M1}+C_{gd,P}\right)+\right. \\ & \left. + \frac{1}{A_0}R_{GB,NMOS}C_{gg,NMOS}\right] + R_3R_{out,D-FSP}R_{M2}\cdot\frac{R_{LOAD}}{R_{LOAD}+r_{o,P}}C_3C_{M2}C_{M1} \end{aligned} \tag{5.18}$$

$$\begin{aligned} b_4 = {} & g_{m,P}R_3R_{out,D-FSP}(R_{LOAD}\|r_{o,P})R_{M2}C_3C_{M2}\left[\left[R_{GB,NMOS}C_{gg,NMOS}\left[\frac{C_{M1}}{A_0}\left[1+\frac{1}{g_{m,CN}r_{o,CN}}\right]\right.\right.\right. \\ & \left.\left.\left. + \frac{C_{M1}+C_{gd,p}}{g_{m,CN}R_{M2}A_0}\right] + \frac{R_{GB,PMOS}C_{M1}C_{gg,PMOS}}{A_1}\right] + \frac{C_{M1}}{g_{m,P}}\left[C_{LOAD}+\frac{R_{GB,NMOS}C_{gg,NMOS}}{A_0g_{m,CN}r_{o,P}R_{M2}}\right]\right] \end{aligned} \tag{5.19}$$

$$\begin{aligned} b_5 = {} & \frac{1}{A_0}R_3R_{out,D-FSP}R_{LOAD}\|r_{o,P}R_{GB,NMOS}C_{M2}C_3C_{gg,NMOS}\cdot \\ & \cdot\left[g_{m,P}C_{M1}\left[\frac{R_{gb,PMOS}C_{gg,PMOS}}{A_1}\left[\frac{1}{g_{m,CN}}+R_{M2}\right]+\frac{R_{M1}C_{gd,P}}{g_{m,CN}}\right]+\right. \\ & \left. + C_{M1}C_{LOAD}\left[\frac{1}{g_{m,CN}}+R_{M2}\right]+\frac{C_{gs,P}C_{LOAD}}{g_{m,CN}}\right] \end{aligned} \tag{5.20}$$

$$\begin{aligned} b_6 = {} & \frac{1}{A_0}R_3R_{out,D-FSP}R_{GB,NMOS}R_{LOAD}\|r_{o,P}C_3C_{M2}C_{gg,NMOS}\cdot \\ & \cdot\left[\frac{1}{A_1}R_{GB,PMOS}C_{gg,PMOS}\left[g_{m,P}R_{M1}C_{gd,P}C_{M1}\left[\frac{1}{g_{m,CN}}+R_{M2}\right]+\right.\right. \end{aligned}$$

$$+\frac{C_{LOAD}}{g_{m,CN}}\left[C_{gs,P}+C_{M1}\right]+\frac{R_{M1}C_{gs,P}C_{M1}}{g_{m,CN}R_{LOAD}\|r_{o,P}}+R_{M2}C_{M1}C_{LOAD}\Bigg]+$$
$$+R_{M1}C_{M1}C_{gs,P}\left[\frac{R_{M2}C_{gd,P}}{g_{m,8}r_{o,8}A_1}+\frac{C_{LOAD}}{g_{m,CN}}\right]\Bigg] \tag{5.21}$$

$$b_7=\frac{1}{A_1A_0}R_3R_{M1}R_{out,D-FSP}R_{GB,NMOS}R_{GB,PMOS}\cdot$$
$$\cdot R_{LOAD}\|r_{o,P}C_{M1}C_{M2}C_3C_{LOAD}C_{gg,NMOS}C_{gg,PMOS}\cdot$$
$$\cdot\left[\frac{C_{gs,P}}{g_{m,CN}}+C_{gd,P}\left(\frac{1}{g_{m,CN}}+R_{M2}\right)+C_{gs,P}\left(R_{M2}+\frac{1}{g_{m,8}g_{m,CN}r_{o,7}}\right)\right] \tag{5.22}$$

The dominant pole is given by Eq. 5.23. $R_{out,D-FSP}$ is the output impedance of the D-FSP block. The non-dominant pole is fixed by the output resistance and C_{LOAD}.

$$\omega_{p1}\approx\frac{1}{g_{m,P}R_{out,D-FSP}(R_{LOAD}\|r_{o,P})\left(C_{M1}+C_{M2}+C_{gd,P}\right)} \tag{5.23}$$

A reduction in the load current, I_{LOAD}, will bring the non-dominant pole closer to the Unity Gain Frequency (UGF), thus degrading the stability. This behaviour is represented by the simplified pole-zero plot in Fig. 5.19. For the sake of clarity, only poles and zeros below 100 MHz are included in the figure. NMC, consisting of components R_{M1}, C_{M1}, R_{M2} and C_{M2}, was used to achieve a proper phase margin in an output current range of 0.1 mA to 100 mA. This resulted in R_{M1} = 1 kΩ, C_{M1} = 5 pF, R_{M2} = 10 kΩ, and C_{M2} = 8 pF.

Post-layout simulations of the open-loop gain are shown in Fig. 5.20 at different load conditions (100 μA, 1 mA, 10 mA and 100 mA). Table 5.4 summarizes the simulated post-layout gain and phase margin values. In every case, the load capacitor is 100 pF, which is the worst-case scenario. Note that the proposed LDO regulator is stable across the entire operating range.

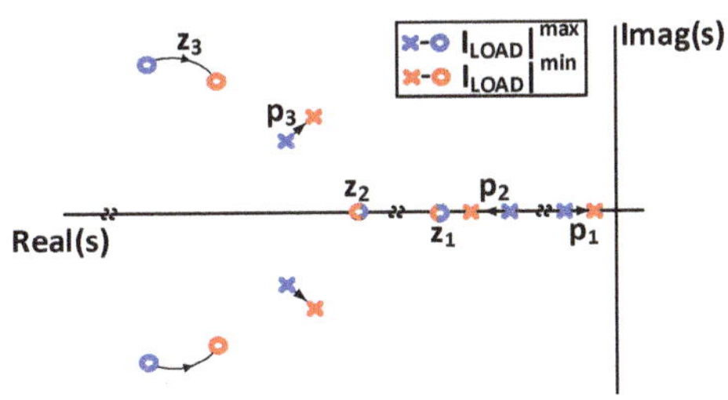

Fig. 5.19 Simplified pole-zero diagram of poles and zeros below 100 MHz

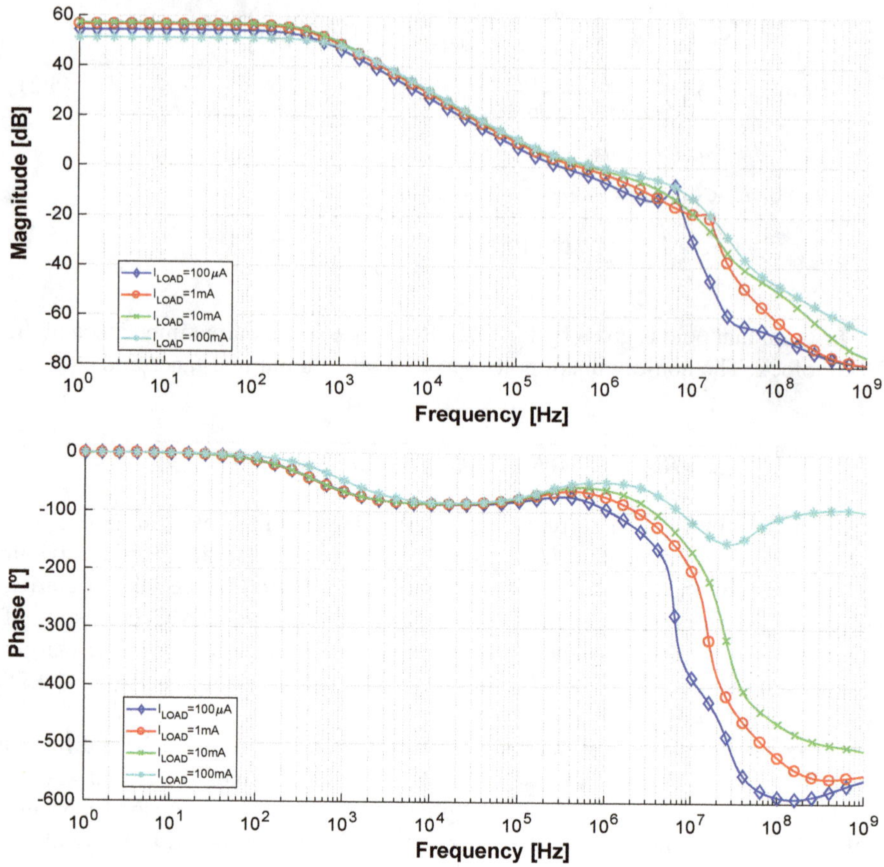

Fig. 5.20 Simulated post-layout open-loop gain of circuit in Fig. 5.14 for $C_{LOAD} = 100$ pF and $V_{IN} = 0.9$ V

Table 5.4 Simulated post-layout gain and phase margin values for different load conditions

I_{LOAD} [mA]	Gain [dB]	PM [°]
100	51.30	132.0
10	57.24	122.2
1	56.65	115.7
0.1	54.38	107.2

5.3.3 Experimental Results

The proposed LDO regulator was designed and implemented in a standard 65-nm CMOS technology. Figure 5.21 shows the layout of the IC-LDO regulator superimposed on a chip micrograph next to the layout of the circuit, where the area denoted by *A* indicates the pass transistor M_{PASS}, and *B* is the core of the circuit including the fast settling blocks. *C* corresponds to the overshoot limiter implemented by M_4, R_2 and C_3, whereas *D* is associated with the biasing circuit. The total area is 340.8 × 135.5 μm. The core of the circuit occupies an area of 90.2 × 135.5 μm. The regulator was designed to drive a maximum load current of 100 mA with a variable C_{LOAD} in the range 0–100 pF.

Experimental results were obtained with the test setup of Appendix B for the worst-case value of C_{LOAD}. Figure 5.22 depicts the line transient response for $V_{OUT} = 0.7$ V and $I_{LOAD} = 100$ mA, with V_{IN} changing from 0.9 to 1.2 V and vice versa. In both cases, the rise and fall times of V_{IN} are 1 μs. Under these conditions, the output voltage shows an overshoot of 45.2 mV and an undershoot of −61.4 mV.

In addition, the worst settling time, which was calculated as the time that takes V_{OUT} to remain inside the 1% error band around its nominal value, is 5.17 μs. Figure 5.23 shows the load transient response for rise and fall times of 1 μs when $V_{IN} = 0.9$ V and the load current changes from 0.1 to 100 mA and vice versa. The voltage V_{OUT} shows a maximum variation of +75.9 mV/−67.5 mV with respect to the nominal voltage $V_{OUT} = 0.7$ V. Under these conditions, the worst settling time is 4.64 μs.

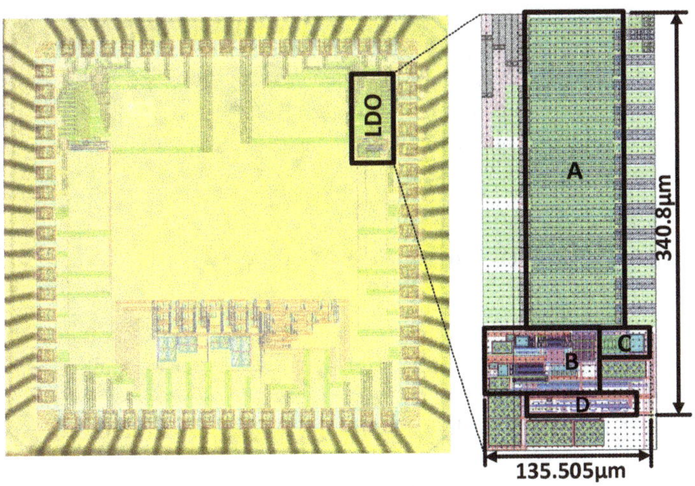

Fig. 5.21 LDO regulator layout superimposed on a chip micrograph

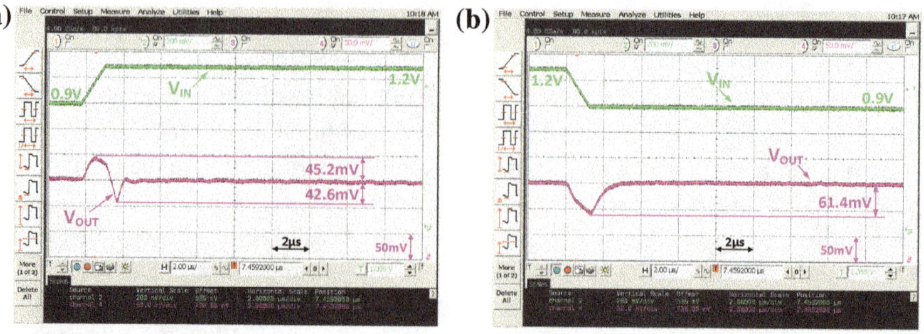

Fig. 5.22 Measured line response with C_{LOAD} = 100 pF and I_{LOAD} = 100 mA. For V_{IN} changing: **a** from 0.9 to 1.2 V, and **b** from 1.2 to 0.9 V, with rise and fall times of 1 μs

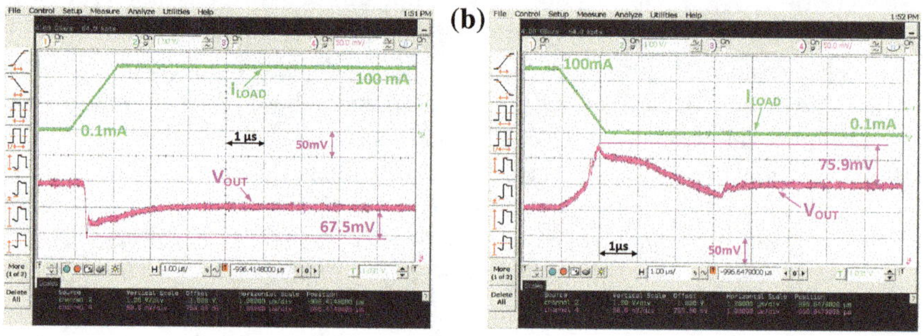

Fig. 5.23 Measured load transient response with C_{LOAD} = 100 pF and V_{IN} = 0.9 V. For I_{LOAD} changing: **a** from 0.1 to 100 mA, and **b** from 100 to 0.1 mA, with rise and fall times of 1 μs

From Figs. 5.22 and 5.23, it can be deduced that the proposed architecture exhibits a fast transient response for changes in both V_{IN} and I_{LOAD}. Table 5.5 summarizes the performances of the proposed LDO regulator.

5.4 Comparison with the State of the Art

Table 5.6 compares the performances of the proposed implementation to those of other LDO regulators published in the literature. The experimental results presented here correspond to the worst-case scenario (according to their respective authors), measured when I_{LOAD} and V_{IN} change between their extreme values. In order to compare different LDO regulators, we use the FOM_1 defined Eq. 2.24 [87]. Figure 5.24 depicts the FOM_1 for different LDO regulators based on the FVF cell (and on derived cells, such as the CAFVF one), which shows that the proposed regulator is close to the state of the art.

Table 5.5 A summary of the performances of the proposed LDO regulator

		Proposed LDO			Proposed LDO
Process	[μm]	65	ΔV_{OUT} varying V_{IN}		
V_{IN}	[V]	0.9–1.2	• Maximum	$[mV]$	45.2
V_{OUT}	[V]	0.7	• Minimum	$[mV]$	−61.4
$V_{DROPOUT}$	[mV]	200	$\Delta V_{IN}/t_r^b$	$[V/\mu s]$	0.3/1
$I_{LOAD,max}$	[mA]	100	ΔV_{OUT} varying I_{LOAD}		
I_Q^a	[μA]	17.38–17.88	• Maximum	$[mV]$	75.9
C_{LOAD}	[pF]	100	• Minimum	$[mV]$	−67.5
$\eta\vert I_{LOAD,max}$	[%]	99.982	$\Delta I_{LOAD}/t_r^b$	$[mA/\mu s]$	9.9/1
Area	$[mm^2]$	0.029	Line regulation	$[mV/V]$	5.61
Response time[a]	$[\mu s]$	5.17	Load regulation	$[\mu V/mA]$	433.8

[a]Worst case, [b]t_r: Rise time

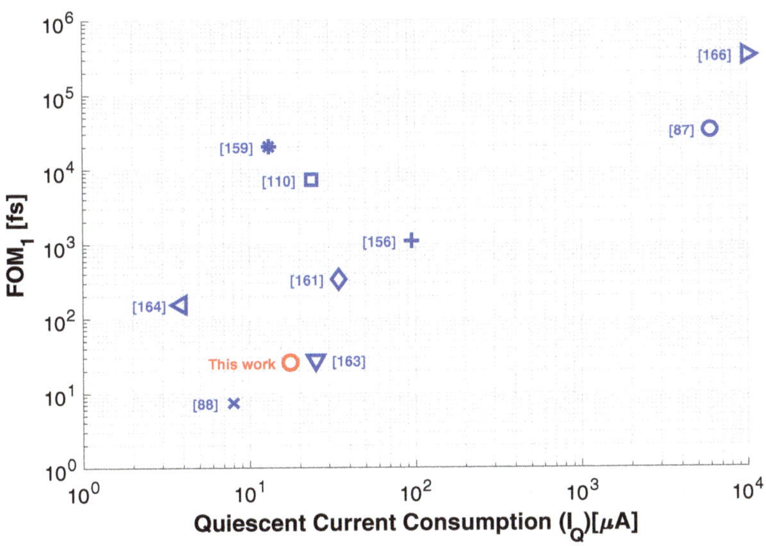

Fig. 5.24 FOM$_1$ versus quiescent current consumption for those IC-LDO regulators based on the FVF cell that perform the best FOM$_1$ (Table 5.6)

Figure 5.25 expands the comparison to the LDO regulators that have the best values of FOM$_1$, regardless their implementation. For the sake of completeness, the regulators designed in Chap. 3 (Chap. 3*), Chap. 4 (Chap. 4) and this chapter (Chap. 5) have been also included in this figure. The FOM$_1$ for the regulator in Chap. 3 has

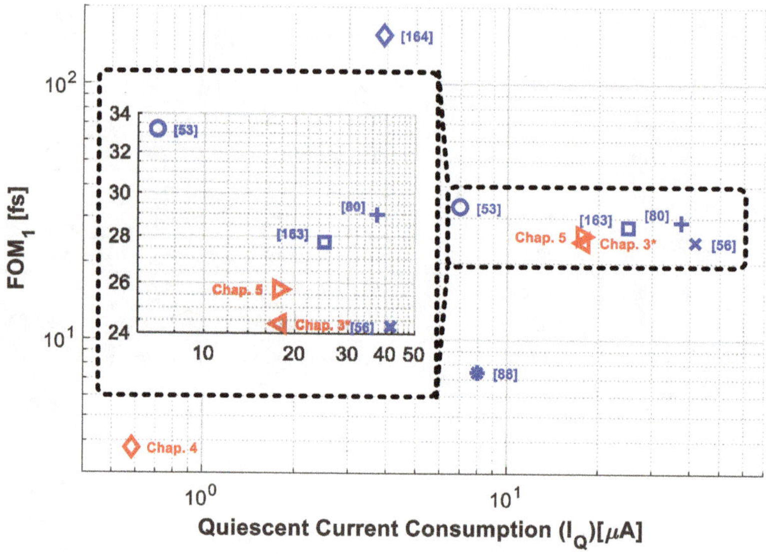

Fig. 5.25 FOM_1 versus quiescent current consumption for those IC-LDO regulators that perform the best FOM_1. Regulators presented in Chap. 3, Chap. 4 and in this chapter (Chap. 5) have been included in the comparison

been computed in the same conditions of reference [30]. It can be seen that the ultra-low-power regulator of Chap. 4 clearly outperforms the rest of regulators published in the literature, concerning FOM_1. As it was stated before, this FOM favours low power consumption. In addition, only a few regulators perform better than those presented in Chap. 3 and in this chapter.

5.5 Conclusions of This Chapter

This chapter has shown that the FVF cell, due to local feedback, is an adequate building block for IC-LDO regulators with low power consumption. However, the basic FVF cell has a low gain and does not implement any mechanism to reduce the negative effect of the input voltage variations. Even more, the charge of the parasitic gate capacitance is limited by the bias current.

Table 5.6 Comparison of the proposed LDO regulator to some of those that have reported the best value of the FOM_1

		[53]	[56]	[80]	[88]	[163]	[164][a]	This work
Process	[μm]	0.35	0.11	0.13	0.09	0.35	0.18	0.065
V_{IN}	[V]	2.5–4.0	1.8–3.8	1.15–1.4	0.75–1.2	1.28–3.3	0.9–1.8	0.9–1.2
V_{OUT}	[V]	2.35	1.6–3.6	1	0.5–1.0	1.1	0.7	0.7
$V_{DROPOUT}$	[mV]	150	200	150	200	180	200	200
$I_{LOAD,max}$	[mA]	100	200	50	100	100	50	100
I_Q[a]	[μA]	7	41.5	37	8	25	3.9	17.38–17.88
C_{LOAD}	[pF]	1e2	4e1	2e1	1e2	1e2	5e1	1e2
$\eta\|I_{LOAD,max}$	[%]	99.993	99.979	99.926	99.992	99.975	99.996	99.982
Area	[mm^2]	0.064	0.21	0.018	0.019	0.126	0.041	0.029
Response time[a]	[μs]	0.15	0.65	0.4	3.75[c]	1.4	_[b]	5.17
ΔV_{OUT} varying V_{IN}								
• Maximum	[mV]	196	_[b]	_[b]	40	20	_[b]	45.2
• Minimum	[mV]	−183	_[b]	_[b]	−33	0	_[b]	−61.4
$\Delta V_{IN}/t_r$[d]	[$V/\mu s$]	0.5/0.5	_[b]	_[b]	0.42/10	1/1e3	_[b]	0.3/1
ΔV_{OUT} varying I_{LOAD}								
• Maximum	[mV]	231	200	56	114	31	500	75.9
• Minimum	[mV]	−243	−385	−42	−73	−80	−500	−67.5
$\Delta I_{LOAD}/t_r$[d]	[$mA/\mu s$]	99.95/0.5	199.5/0.5	49.95/0.2	98.5/0.1	100/0.5	50/0.1	99.9/1
Line regulation	[mV/V]	1	8.9	8.1	3.78	_[b]	_[b]	5.61
Load regulation	[$\mu V/mA$]	80.00	108.0	55.6	100.0	190.0	_[b]	433.8
FOM_1	[fs]	33.18	24.28	29.01	7.48	27.75	156.0	25.64

[a]Worst case, [b]Not available, [c]Estimation based on published results, [d]t_r: Rise time

To solve these issues, a new CAFVF-based IC-LDO regulator has been proposed. This cell introduces a gain block to increase the open-loop gain, and consequently, to enhance the load and line regulations. In addition, RC couplings have been included to improve the transient response, not only to load but also, for the first time, to line variations. The proposed regulator has been fabricated in a standard 65-nm CMOS technology, and experimental results show that it is close to the state of the art.

Chapter 6
Conclusions

As a consequence of technology downscaling, the coexistence of complex subsystems in the same chip becomes plausible, leading to the SoC paradigm. However, this technological evolution entails an increase in the complexity of power management systems, as different voltage supply domains, with different requirements of voltage level, ripple, maximum capacitive load and maximum output current, coexist in the same chip.

Voltage regulators are key elements in power management systems. Chapter 1 reviews the most common implementations of voltage regulators, with emphasis on linear regulators, which are preferred when a fast transient response, low ripple and good power supply ripple rejection are required. To attain the efficiency of switching regulators, LDO regulators are linear cells with a low dropout voltage in the pass transistor. Usually, LDO regulators have stability problems, which can be solved by forcing a dominant pole at the output node. This is achieved by placing a large external capacitor, at the cost of an increase in the PCB complexity and a slow transient response. However, following the present trend of larger system integration, IC-LDO regulators are gaining popularity. In IC-LDO regulators, the dominant pole is located at an internal node, usually at the gate of the pass transistor, taking advantage of the large parasitic capacitance of this node. As no external components are needed, the PCB is simplified and the regulator size is reduced. Moreover, as these regulators can be completely integrated, the power supply distribution is handled inside the same silicon die, leading to higher power efficiency.

Chapter 2 is devoted to the basics of IC-LDO regulators. Line and load regulations are defined and their main design challenges: stability, transient response and power supply ripple rejection are discussed. In every case, a thorough revision of the solutions presented in the literature is made. The chapter concludes with the definition of the most common Figures of Merit (FOMs) that allow a fair comparison between IC-LDO regulators proposed in the literature.

The stability of IC-LDO regulators is the objective of Chap. 3. The analysis of the classical topology for IC-LDO regulators shows that some kind of compensation

© Springer International Publishing AG, part of Springer Nature 2018

J. M. Hinojo et al., *Internally Compensated LDO Regulators for Modern System-on-Chip Design*, https://doi.org/10.1007/978-3-319-75411-6_6

is required; even more, this compensation is highly dependent on the load current, so that the classical Miller compensation is not enough to stabilize the regulator in the entire range of operation. To solve this problem a new Miller-type compensation scheme is proposed where the zero-nulling resistor is adaptively changed as a function of the load current. To this end, a replica circuit determines the operation point. Simulation and experimental results show that this technique is able to guarantee the regulator stability in the entire operation range.

Today, in the IoT era, WSNs are gaining popularity. In most cases, sensors nodes are autonomous systems supplied by batteries which, in some cases, are complemented with energy scavenging devices. To achieve a battery lifetime in the order of years with small-sized batteries, LDO regulators with ultra-low quiescent power consumption (consuming less than 1 μW) are required. Chapter 4 discusses the challenges related to the design of ultra-low quiescent power IC-LDO regulators and presents an innovative solution based on the classical topology that replaces the output stage of the error amplifier with a highly efficient class AB buffer that drives the large parasitic capacitance at the gate of the pass transistor.

Finally, Chap. 5 deals with IC-LDO topologies other than the classical one. In particular, it focuses on those regulators based on the Flipped Voltage Follower (FVF) family of cells. These regulators benefit from the good performances of these cells regarding low output impedance, class AB behaviour and high-frequency stability. Chapter 5 makes a complete review of the solutions based on the FVF cell that have been presented in the literature and proposes a new regulator based on the CAFVF cell that includes fast charging/discharging paths for the gate capacitance of the pass transistor, and RC couplings to improve the transient response to load and line variations.

The IC-LDO regulators proposed in Chaps. 3, 4 and 5 have been fabricated in standard CMOS technologies (65-nm for regulators in Chaps. 3 and 5, and 180-nm in Chap. 4). In every case, measured performances are compared to those obtained with the best IC-LDO regulators presented in the literature using the FOMs defined in Chap. 2. These comparisons show that the proposed regulators are in, or close to, the state of the art.

Appendix A
Notation

In this appendix, the general notation used in this book is defined.

In general throughout the book an upper-case letter with an upper-case subscript is used to denote an electrical magnitude, regardless its DC, small-signal or instantaneous value. On the other hand, on those specific cases where it is necessary to distinguish between the DC component, small-signal and instantaneous value the following convention is chosen:

- **DC component**: It is represented by an upper-case letter with an upper-case subscript. Example: V_{OUT}, V_{IN} or I_{LOAD}.
- **Deviations with respect to the DC component**: It is represented by a lower case letter with a lower case subscript. Example: v_{out}, v_{in} or i_{load}. Despite these deviations can have a small or large value, the symbol is normally used to represent the small-signal component of the corresponding variable.
- **Instantaneous value**: It is represented by a lower case letter with an upper-case subscript. Example: v_{OUT}, v_{IN} or i_{LOAD}.

According to the definitions above:

$$\begin{aligned} v_{OUT} &= V_{OUT} + v_{out} \\ v_{IN} &= V_{IN} + v_{in} \\ i_{LOAD} &= I_{LOAD} + i_{load} \end{aligned}$$

The most common abbreviations that appear in this book are now introduced.

- $\mathbf{A_{OL}}$: open-loop gain.
- $\boldsymbol{\omega_{pi}}$: i-th pole of the transfer function, in ascending order from the lowest frequency (dominant pole) onwards.
- $\boldsymbol{\omega_{zi}}$: i-th zero of the transfer frequency, in ascending order from the lowest frequency onwards.
- $\mathbf{R_{LOAD}}$: resistance that models the load current, $\mathrm{I_{LOAD}}$. It is estimated as $R_{LOAD} = V_{OUT}/I_{LOAD}$.

© Springer International Publishing AG, part of Springer Nature 2018

J. M. Hinojo et al., *Internally Compensated LDO Regulators for Modern System-on-Chip Design*, https://doi.org/10.1007/978-3-319-75411-6

- $\mathbf{R_{OUT}}$: equivalent output resistance (including R_{LOAD}) at the IC-LDO regulator output.
- $\mathbf{C_{LOAD}}$: load capacitor located at the output node of an LDO regulator.
- $\mathbf{C_{par}}$: parasitic capacitance at the output of the LDO regulator.
- $\mathbf{C_{OUT}}$: total capacitance at the output of the LDO regulator. $C_{OUT} = C_{LOAD} + C_{par}$. Usually, C_{par} is neglected when compared to C_{LOAD}, so that $C_{OUT} \approx C_{LOAD}$.
- $\mathbf{Z_{OUT}}$: total impedance at the regulator output, formed by the parallel arrangement of R_{OUT} and C_{OUT}.
- $\mathbf{g_{m,EA}}$: transconductance of the error amplifier.
- $\mathbf{R_{O,EA}}$: output resistance of the error amplifier.
- $\mathbf{C_{O,EA}}$: output capacitance at the output of the error amplifier.
- $\mathbf{g_{m,P}}$: small-signal transconductance of M_{PASS}.
- $\mathbf{r_{o,P}}$: small-signal resistance of M_{PASS}.
- $\mathbf{C_{gs,P}}$, $\mathbf{C_{gd,P}}$, $\mathbf{C_{gb,P}}$, $\mathbf{C_{db,P}}$: gate-to-source, gate-to-drain, gate-to-bulk and drain-to-bulk small-signal parasitic capacitances of M_{PASS}.
- $\mathbf{g_{m,i}}$: small-signal transconductance of transistor M_i.
- $\mathbf{r_{o,i}}$: small-signal resistance of transistor M_i.
- $\mathbf{C_{gs,i}}$, $\mathbf{C_{gd,i}}$, $\mathbf{C_{gb,i}}$, $\mathbf{C_{db,i}}$: gate-to-source, gate-to drain, gate-to-bulk and drain-to-bulk small-signal parasitic capacitances of transistor M_i.
- $\mathbf{C_{GATE}}$: the parasitic capacitance, referenced to ground, at M_{PASS} gate. This capacitance includes the total gate parasitic capacitance of M_{PASS} and the parasitic capacitance at the output of the error amplifier.
- $\mathbf{C_C}$: compensation resistor (in Miller-compensated regulators).
- $\mathbf{R_C}$: zero-nulling resistor (in Miller-compensated regulators).
- $\mathbf{R_{ON}}$: on resistance of switches.

Appendix B
Test Setup

This appendix describes the setup used to carry out the measurements presented in this book. This is a list of the equipment:

- **Oscilloscope**: Agilent MSO8104A
- **Signal Generator**: Rohde & Schwarz AM300
- **Digital Multimeter (DMM)**: HP 34401A
- **Bench DC Power Supply**: Agilent U3630A
- **Vector Network Analyzer (VNA)**: Rohde & Schwarz ZVRL
- **Spectrum Analyzer**: Rohde & Schwarz FSU

B.1 Measuring the Transient Response to an Input Voltage Step

The line transient response can be measured using the setup shown in Fig. B.1, if an adequate Bias Tee is available. Otherwise, it is possible to use a voltage buffer, e.g. LT1210 from Linear Technologies®, which is able to provide enough current to the LDO regulator under test, Fig. B.2. In that case, the input voltage signal was varied from $V_{IN,min}$ to $V_{IN,max}$. Note that a resistor R_{ADAPT} is included to properly adapt the signal generator input impedance, typically 50 Ω.

B.2 Measuring the Transient Response to a Step Variation of the Load Current

The load transient response has been measured using the setup shown in Fig. B.3.

© Springer International Publishing AG, part of Springer Nature 2018

J. M. Hinojo et al., *Internally Compensated LDO Regulators for Modern System-on-Chip Design*, https://doi.org/10.1007/978-3-319-75411-6

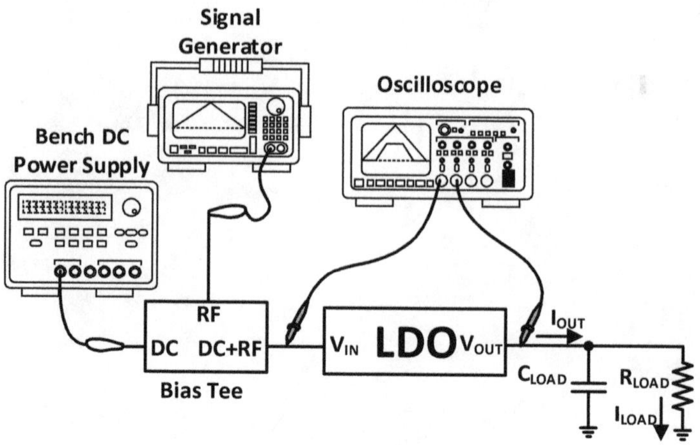

Fig. B.1 Line transient response measurement setup using a Bias Tee

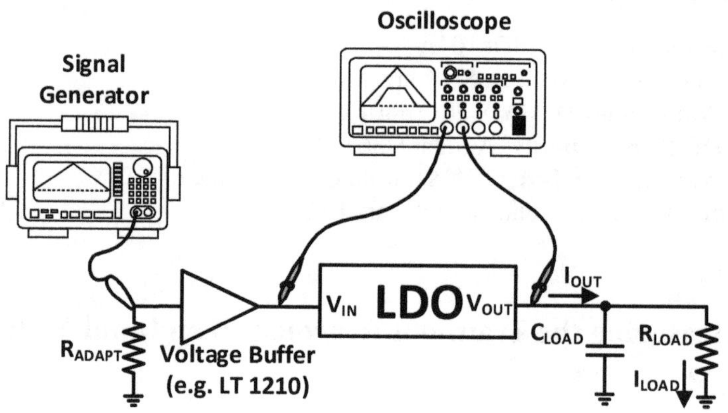

Fig. B.2 Line transient response measurement setup using a voltage buffer

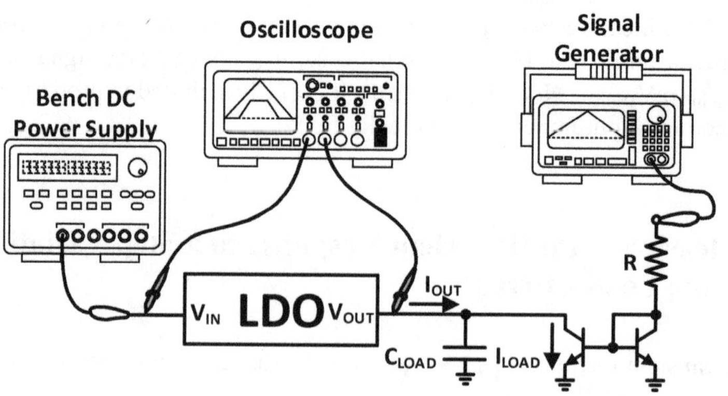

Fig. B.3 Load transient response measurement setup

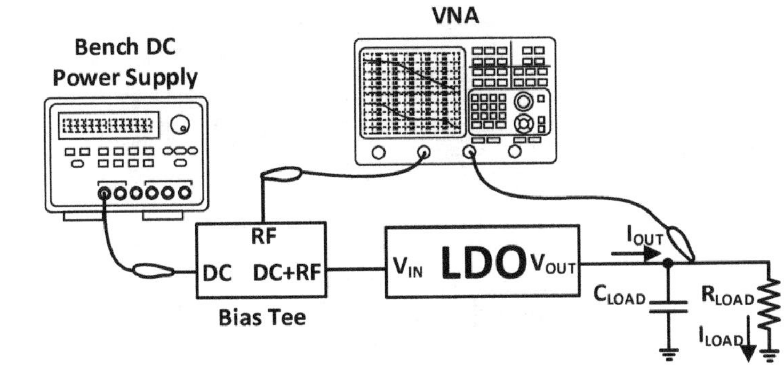

Fig. B.4 PSRR measurement setup using a Bias Tee

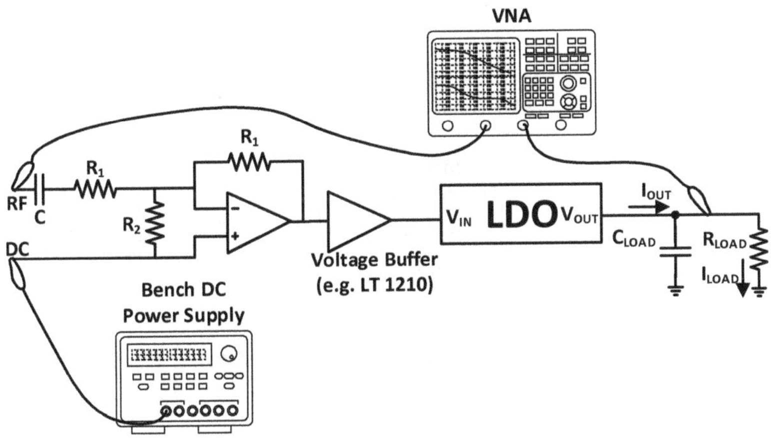

Fig. B.5 PSRR measurement setup using a voltage buffer

B.3 Measuring the PSRR

Figure B.4 shows the PSRR measurement setup when an adequate Bias Tee and VNA are available.

In those cases where it is not possible to use a Bias Tee to inject the AC signal over the DC level, the setup shown in Fig. B.5 can be used.

Finally, if the equipment used in the measurements of Figs. B.4 and B.5 is unavailable, the alternative measurement setup proposed in [83] can be used, Fig. B.6. The main disadvantage of this setup is that two spectrum analysers are required.

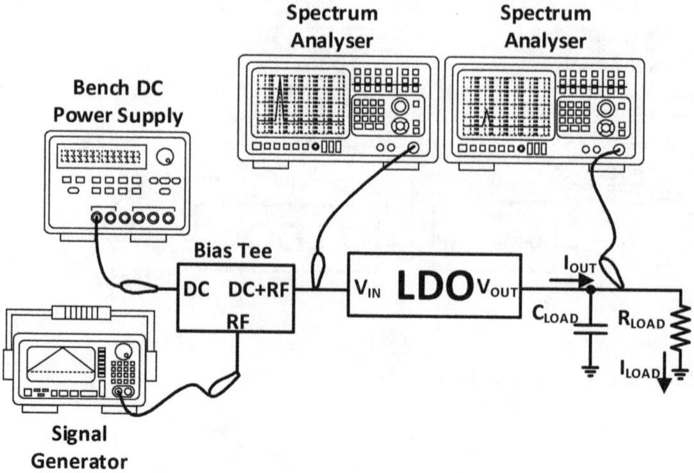

Fig. B.6 PSRR measurement setup using two spectrum analysers

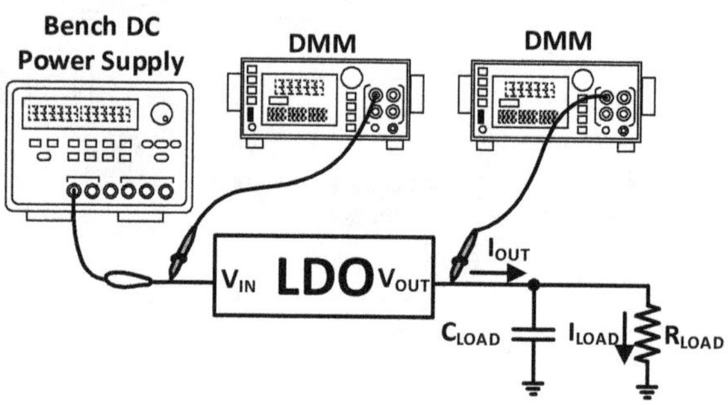

Fig. B.7 Line regulation measurement setup

B.4 Measuring the Line and Load Regulations

Figures B.7 and B.8 show a setup to evaluate the line and load regulations, respectively. In both cases, the load current can be implemented by a resistor or an electronic load.

In addition, note that to evaluate the load regulation, it is necessary to insert an ammeter at the output node in order to register the current.

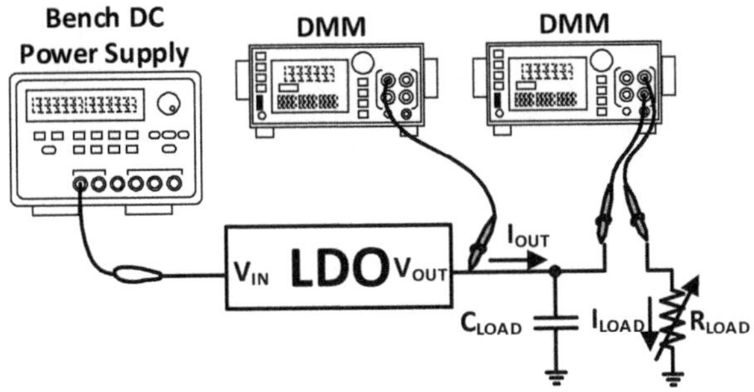

Fig. B.8 Load regulation measurement setup

Appendix C
Some Considerations for Design of the Prototype Test Boards

General layout guidelines for printed circuit boards (PCB) must be taken into account when designing a test board for IC-LDO regulators. Some of the most common layout techniques for low-noise and accurate measurements recommendations are

- preserving the integrity of the signals,
- minimizing length of PCB tracks to reduce their parasitic inductance and, hence, the voltage drops,
- using decoupling capacitor as close as possible to the power supply pins to reduce the inductive effect of tracks,
- reducing external noise using shielding cables and special connectors, e.g. tri-axial cables,
- avoiding to cut ground planes to reduce electromagnetic interferences, or
- using ferrite beads to filter as much as possible the power supply noise.

Following these recommendations, test boards designed to evaluate the performances of IC-LDO regulators presented in this book are described here.

C.1 PCB for IC-LDO Presented in Chap. 3

Figure C.1 shows the test PCB designed using a two-layer FR4 board. As it can be observed, the internal power supply generation circuitry is located close to the IC-LDO under test to minimize the length of PCB tracks. In addition, an ultra-fast linear regulator with a high PSRR has been chosen in order to minimize the noise effect and voltage drops due to external sources (e.g. avoiding the effect of the bench DC power supply regulation) in the measurements. Some test points (TPx in Fig. C.1) have also been included as close as possible to the relevant pins, e.g. the output voltage, to reduce the inductive effect of the oscilloscope probe.

Note that, as far as possible, the ground plane covers the entire PCB, and it is only cut in those areas where no relevant tracks are routed.

© Springer International Publishing AG, part of Springer Nature 2018

J. M. Hinojo et al., *Internally Compensated LDO Regulators for Modern System-on-Chip Design*, https://doi.org/10.1007/978-3-319-75411-6

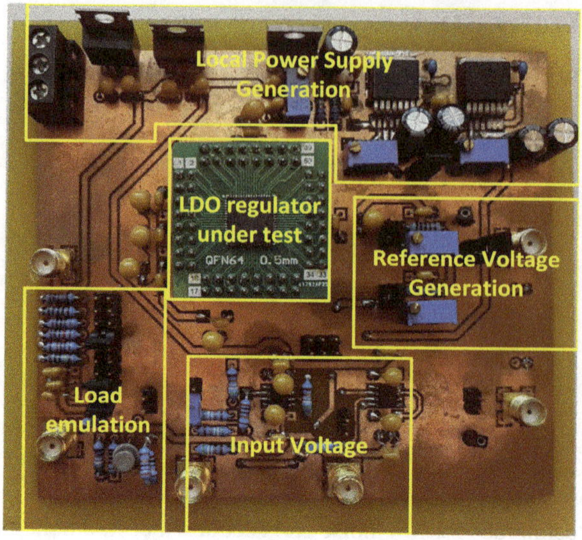

Fig. C.1 Photograph of the PCB designed to test the LDO regulator proposed in Chap. 3

C.2 PCB for IC-LDO Presented in Chap. 4

Figure C.2 shows the test PCB designed using a two-layer FR4 board. In this figure, local power supply is located at the top of the figure and it is generated by means of an ultra-fast linear regulator with high PSRR. In addition, some Test Points (TP_x) in the figure have been laid out on the board as close as possible to the most relevant pins of the test chip, like the output voltage pin. Note that, as far as possible, the ground plane covers the entire PCB, and plane cuts are restricted to those zones where no relevant tracks are routed.

C.3 PCB for IC-LDO Presented in Chap. 5

The PCB used to test the FVF-based regulator is shown in Fig. C.3. In this case, the chip was designed with several biasing currents and reference voltages which have to be generated in the prototype board. Concretely, for bias current generation, guard rings for the inputs of the selected opamps were implemented to minimize noise and leakage at their inputs terminal. In addition, due to the size of the board (around 20 cm x 20 cm), different decoupling capacitors were used for each discrete opamp. An adapter was used to test the proposed regulator, as it was delivered in a Surface-Mount Technology package. To reduce parasitic inductance and resistance, the length of its tracks was minimized.

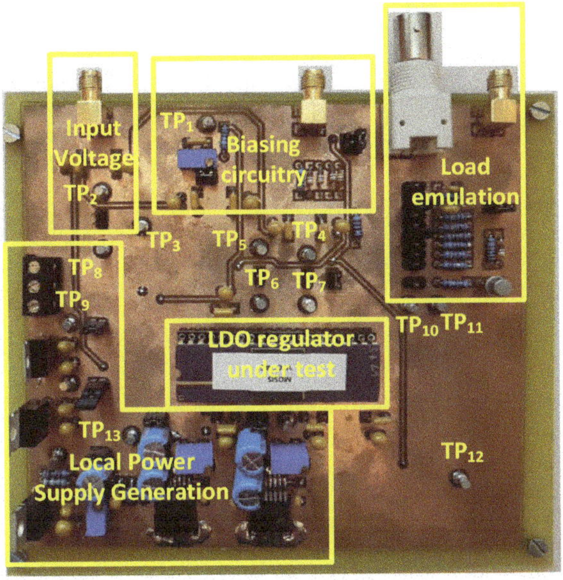

Fig. C.2 Photograph of the PCB designed to test the LDO regulator proposed in Chap. 4

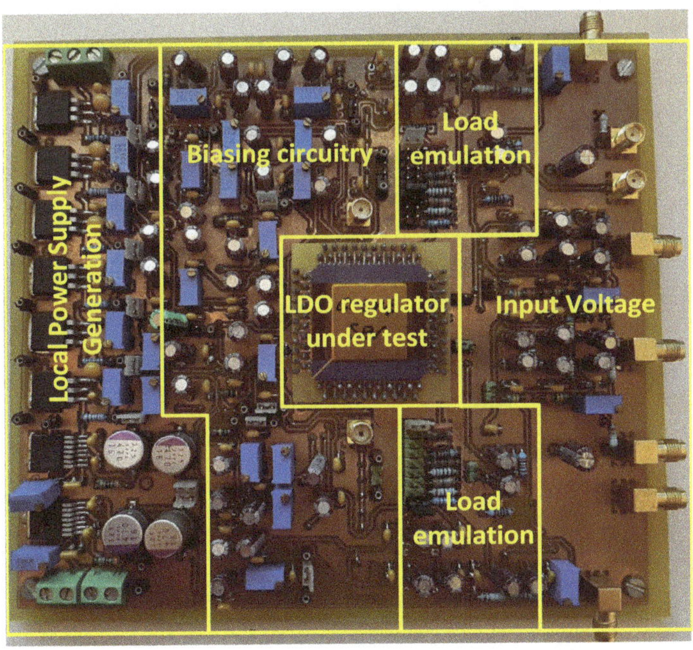

Fig. C.3 Photograph of the PCB designed to test the LDO regulator proposed in Chap. 5

References

1. A. Zanella, N. Bui, A. Castellani, L. Vangelista, M. Zorzi, Internet of things for smart cities. IEEE Internet Things J. **1**(1), 22–32 (2014)
2. J. Jin, J. Gubbi, S. Marusic, M. Palaniswami, An information framework for creating a smart city through internet of things. IEEE Internet Things J. **1**(2), 112–121 (2014)
3. P. Vlacheas, R. Giaffreda, V. Stavroulaki, D. Kelaidonis, V. Foteinos, G. Poulios, P. Demestichas, A. Somov, A. Biswas, K. Moessner, Enabling smart cities through a cognitive management framework for the internet of things. IEEE Commun. Mag. **51**(6), 102–111 (2013)
4. S.R Islam, D. Kwak, M.H. Kabir, M. Hossain, K.-S. Kwak, The internet of things for health care: a comprehensive survey. IEEE Access **3**, 678–708 (2015)
5. Farsens (2015), http://www.farsens.com/
6. Enocean (2015), https://www.enocean.com/
7. Panstamp (2015), http://www.panstamp.com/
8. Libelium (2015), http://www.libelium.com/
9. S. Mann, Wearable computing: a first step toward personal imaging. Computer **30**(2), 25–32 (1997)
10. G.E. Moore, Cramming more components onto integrated circuits. Electronics **38**(8) (1965)
11. ITRS, International technology roadmap for semiconductors 2013 edition executive summary (2013), http://www.itrs2.net/2013-itrs.html
12. G. Hurtz, D. Sugasawara, Cdma handset design challenge: 11 separate power supplies (2003), http://www.eetimes.com/document.asp?doc_id=1226477
13. J. Dickson, On-chip high-voltage generation in mnos integrated circuits using an improved voltage multiplier technique. IEEE J. Solid-State Circuits **11**(3), 374–378 (1976)
14. R. Pelliconi, D. Iezzi, A. Baroni, M. Pasotti, P. Rolandi, Power efficient charge pump in deep submicron standard cmos technology. IEEE J. Solid-State Circuits **38**(6), 1068–1071 (2003)
15. F. Su, W.-H. Ki, Design strategy for step-up charge pumps with variable integer conversion ratios. IEEE Trans. Circuits Syst. II: Express Br. **54**(5), 417–421 (2007)
16. J. Starzyk, Y.-W. Jan, F. Qiu, A dc-dc charge pump design based on voltage doublers. IEEE Trans. Circuits Syst. I: Fundam. Theory Appl. **48**(3), 350–359 (2001)
17. M. AbdElFattah, A. Mohieldin, A. Emira, E. Sanchez-Sinencio, A low-voltage charge pump for micro scale thermal energy harvesting, in *IEEE International Symposium on Industrial Electronics (ISIE)* (2011), pp. 76–80
18. G. Palumbo, D. Pappalardo, Charge pump circuits: an overview on design strategies and topologies. IEEE Circuits Syst. Mag. **10**(1), 31–45 (2010)
19. Y. Ramadass, A. Chandrakasan, Voltage scalable switched capacitor dc-dc converter for ultra-low-power on-chip applications, in *IEEE Power Electronics Specialists Conference, PESC* (2007), pp. 2353–2359
20. V. Ng, S. Sanders, A 92 switched-capacitor dc-dc converter, in *IEEE International Solid-State Circuits Conference Digest of Technical Papers (ISSCC)* (2012), pp. 282–284

© Springer International Publishing AG, part of Springer Nature 2018

J. M. Hinojo et al., *Internally Compensated LDO Regulators for Modern System-on-Chip Design*, https://doi.org/10.1007/978-3-319-75411-6

21. A.P. Chandrakasan, R.W. Brodersen, *Low Power Digital CMOS Design* (Kluwer Academic Publishers, Norwell, MA, USA, 1995)
22. A. Stratakos, S. Sanders, R. Brodersen, A low-voltage cmos dc-dc converter for a portable battery-operated system, in *25th Annual IEEE Power Electronics Specialists Conference, PESC*, vol.1 (1994), pp. 619–626
23. Y. Ramadass, A. Chandrakasan, Minimum energy tracking loop with embedded dc-dc converter enabling ultra-low-voltage operation down to 250 mv in 65 nm cmos. IEEE J. Solid-State Circuits **43**(1), 256–265 (2008)
24. E. Soenen, A. Roth, J. Shi, M. Kinyua, J. Gaither, E. Ortynska, A robust digital dc-dc converter with rail-to-rail output range in 40 nm cmos," in *IEEE International Solid-State Circuits Conference Digest of Technical Papers (ISSCC)* (2010), pp. 198–199
25. S. Bandyopadhyay, Y. Ramadass, A. Chandrakasan, 20 μa to 100 ma dc-dc converter with 2.8 to 4.2 v battery supply for portable applications in 45 nm cmos, in *IEEE International of Solid-State Circuits Conference Digest of Technical Papers (ISSCC)* (2011), pp. 386–388
26. Y. Ramadass, A. Chandrakasan, A battery-less thermoelectric energy harvesting interface circuit with 35 mv startup voltage. IEEE J. Solid-State Circuits **46**(1), 333–341 (2011)
27. Y. Ramadass, A. Fayed, B. Haroun, A. Chandrakasan, A 0.16 mm^2 completely on-chip switched-capacitor dc-dc converter using digital capacitance modulation for ldo replacement in 45 nm cmos," in *IEEE International of Solid-State Circuits Conference Digest of Technical Papers (ISSCC)* (2010), pp. 208–209
28. J. Xiao, A. Peterchev, J. Zhang, S. Sanders, A 4 μa quiescent-current dual-mode digitally controlled buck converter ic for cellular phone applications. IEEE J. Solid-State Circuits **39**(12), 2342–2348 (2004)
29. G. Rincon-Mora, *Analog IC Design with Low-Dropout Regulators (LDOs)*, 1st edn. (McGraw-Hill Inc, New York, NY, USA, 2009)
30. K.N. Leung, P. Mok, A capacitor-free cmos low-dropout regulator with damping-factor-control frequency compensation. IEEE J. Solid-State Circuits **38**(10), 1691–1702 (2003)
31. A. Leung, P. Mok, W.H. Ki, J. Sin, Damping-factor-control frequency compensation technique for low-voltage low-power large capacitive load applications, in *IEEE International of Solid-State Circuits Conference, 1999. Digest of Technical Papers. ISSCC* (1999), pp. 158–159
32. K.N. Leung, P. Mok, Analysis of multistage amplifier-frequency compensation. IEEE Trans. Circuits Syst. I: Fundam. Theory Appl. **48**(9), 1041–1056 (2001)
33. S.K. Lau, P. Mok, K.N. Leung, A low-dropout regulator for soc with q -reduction. IEEE J. Solid-State Circuits **42**(3), 658–664 (2007)
34. B. Ahuja, An improved frequency compensation technique for cmos operational amplifiers. IEEE J. Solid-State Circuits **18**(6), 629–633 (1983)
35. C. Chava, J. Silva-Martinez, A frequency compensation scheme for ldo voltage regulators. IEEE Trans. Circuits Syst. I: Regul. Pap. **51**(6), 1041–1050 (2004)
36. G. Rincon-Mora, Active capacitor multiplier in miller-compensated circuits. IEEE J. Solid-State Circuits **35**(1), 26–32 (2000)
37. R. Milliken, J. Silva-Martinez, E. Sanchez-Sinencio, Full on-chip cmos low-dropout voltage regulator. IEEE Trans. Circuits Syst. I: Regul. Pap. **54**(9), 1879–1890 (2007)
38. Pmos low drop-out voltage regulator using non-inverting variable gain stage, U.S. Patent No. 6,246,221 (2001), http://www.google.com/patents/US6246221
39. R. Dokania, G. Rincon-Mora, Cancellation of load regulation in low drop-out regulators. Electron. Lett. **38**(22), 1300–1302 (2002)
40. G. Giustolisi, G. Palumbo, E. Spitale, Ldo compensation strategy based on current buffer/amplifiers, in *18th European Conference on Circuit Theory and Design, ECCTD* (2007), pp. 116–119
41. L.-G. Shen, Z.-S. Yan, X. Zhang, Y.-F. Zhao, A capacitor-less low-dropout regulator for soc with bi-directional asymmetric buffer, in *IEEE International Symposium on Circuits and Systems, ISCAS* (2008), pp. 2677–2680
42. G. Giustolisi, G. Palumbo, E. Spitale, Robust miller compensation with current amplifiers applied to ldo voltage regulators. IEEE Trans. Circuits Syst. I: Regul. Pap. **59**(9), 1880–1893 (2012)

43. A. Garimella, M. Rashid, P. Furth, Frequency compensation techniques using current buffers, in *53rd IEEE International Midwest Symposium on Circuits and Systems (MWSCAS)* (2010), pp. 578–579
44. A. Garimella, M. Rashid, P. Furth, Reverse nested miller compensation using current buffers in a three-stage ldo. IEEE Trans. Circuits Syst. II: Express Br. **57**(4), 250–254 (2010)
45. V. Balan, A low-voltage regulator circuit with self-bias to improve accuracy. IEEE J. Solid-State Circuits **38**(2), 365–368 (2003)
46. Y.-H. Lam, W.-H. Ki, C.-Y. Tsui, Adaptively-biased capacitor-less cmos low dropout regulator with direct current feedback, in *Asia and South Pacific Conference on Design Automation* (2006), p. 2
47. T. Man, P. Mok, M. Chan, A high slew-rate push-pull output amplifier for low-quiescent current low-dropout regulators with transient-response improvement. IEEE Trans. Circuits Syst. II: Express Br. **54**(9), 755–759 (2007)
48. C. Zhan, W.-H. Ki, Output-capacitor-free adaptively biased low-dropout regulator for system-on-chips. IEEE Trans. Circuits Syst. I: Regul. Pap. **57**(5), 1017–1028 (2010)
49. E. Ho, P. Mok, A capacitor-less cmos active feedback low-dropout regulator with slew-rate enhancement for portable on-chip application. IEEE Trans. Circuits Syst. II: Express Br. **57**(2), 80–84 (2010)
50. M. Ho, K.N. Leung, Dynamic bias-current boosting technique for ultralow-power low-dropout regulator in biomedical applications. IEEE Trans. Circuits Syst. II: Express Br. **58**(3), 174–178 (2011)
51. C. Zheng, D. Ma, Design of monolithic cmos ldo regulator with d^2 coupling and adaptive transmission control for adaptive wireless powered bio-implants. IEEE Trans. Circuits Syst. I: Regul. Pap. **58**(10), 2377–2387 (2011)
52. C. Zhan, W.-H. Ki, An output-capacitor-free adaptively biased low-dropout regulator with subthreshold undershoot-reduction for soc. IEEE Trans. Circuits Syst. I: Regul. Pap. **59**(5), 1119–1131 (2012)
53. X. Ming, Q. Li, Z. kun Zhou, B. Zhang, An ultrafast adaptively biased capacitorless ldo with dynamic charging control. IEEE Trans. Circuits Syst. II: Express Br. **59**(1), 40–44 (2012)
54. S. Chong, P.K. Chan, A 0.9-μa quiescent current output-capacitorless ldo regulator with adaptive power transistors in 65-nm cmos. IEEE Trans. Circuits Syst. I: Regul. Pap. **60**(4), 1072–1081 (2013)
55. X. Qu, Z. kun Zhou, B. Zhang, Z.-J. Li, An ultralow-power fast-transient capacitor-free low-dropout regulator with assistant push-pull output stage. IEEE Trans. Circuits Syst. II: Express Br. **60**(2), 96–100 (2013)
56. Y. il Kim, S.-S. Lee, A capacitorless ldo regulator with fast feedback technique and low-quiescent current error amplifier. IEEE Trans. Circuits Syst.II: Express Br. **60**(6), 326–330 (2013)
57. S.S. Chong, P.K. Chan, A sub-1 v transient-enhanced output-capacitorless ldo regulator with push-pull composite power transistor. IEEE Trans. Very Large Scale Integr. (VLSI) Syst. **22**(11), 2297–2306 (2014)
58. X.L. Tan, S.S. Chong, P.K. Chan, U. Dasgupta, A ldo regulator with weighted current feedback technique for 0.47–10 nf capacitive load. IEEE J. Solid-State Circuits **49**(11), 2658–2672 (2014)
59. H. Aminzadeh, M. Nabavi, W. Serdijn, Low-dropout voltage source: an alternative approach for low-dropout voltage regulators. IEEE Trans. Circuits Syst. II: Express Br. **61**(6), 413–417 (2014)
60. K. Keikhosravy, S. Mirabbasi, A 0.13-μm cmos low-power capacitor-less ldo regulator using bulk-modulation technique. IEEE Trans. Circuits Syst. I: Regul. Pap. **61**(11), 3105–3114 (2014)
61. X. Qu, Z. kun Zhou, B. Zhang, Ultralow-power fast-transient output-capacitor-less low-dropout regulator with advanced adaptive biasing circuit. IET Circuits, Devices Syst. **9**(3), 172–180 (2015)

62. A. Maity, A. Patra, Tradeoffs aware design procedure for an adaptively biased capacitorless low dropout regulator using nested miller compensation. IEEE Trans. Power Electron. **31**(1), 369–380 (2016)
63. M. Degrauwe, J. Rijmenants, E. Vittoz, H. De Man, Adaptive biasing cmos amplifiers. IEEE J. Solid-State Circuits **17**(3), 522–528 (1982)
64. A. Lopez-Martin, S. Baswa, J. Ramirez-Angulo, R. Carvajal, Low-voltage super class ab cmos ota cells with very high slew rate and power efficiency. IEEE J. Solid-State Circuits **40**(5), 1068–1077 (2005)
65. J. Roh, High-gain class-ab ota with low quiescent current. Analog Integr. Circuits Signal Process. **47**(2), 225–228 (2006), http://dx.doi.org/10.1007/s10470-006-4959-1
66. J. Ramirez-Angulo, R. Carvajal, A. Torralba, J. Galan, A. Vega-Leal, J. Tombs, The flipped voltage follower: a useful cell for low-voltage low-power circuit design, in *IEEE International Symposium on Circuits and Systems, ISCAS*, vol. 3 (2002), pp. 615–618
67. R. Carvajal, J. Ramirez-Angulo, A. Lopez-Martin, A. Torralba, J. Galan, A. Carlosena, F. Chavero, The flipped voltage follower: a useful cell for low-voltage low-power circuit design. IEEE Trans. Circuits Syst. I: Regul. Pap. **52**(7), 1276–1291 (2005)
68. Y.-H. Lam, W.-H. Ki, A 0.9v 0.35 μm adaptively biased cmos ldo regulator with fast transient response, in *IEEE International Solid-State Circuits Conference, ISSCC. Digest of Technical Papers* (2008), pp. 442–626
69. D. Semiconductor/Maxim, Application note 883: Improved power supply rejection for ic linear regulators (2002), https://www.maximintegrated.com/en/app-notes/index.mvp/id/883
70. V. Gupta, G. Rincon-Mora, A low dropout, cmos regulator with high psr over wideband frequencies, in *IEEE International Symposium on Circuits and Systems, ISCAS*, vol. 5 (2005), pp. 4245–4248
71. V. Gupta, G. Rincon-Mora, A 5ma 0.6 μm cmos miller-compensated ldo regulator with -27db worst-case power-supply rejection using 60pf of on-chip capacitance, in *IEEE International Solid-State Circuits Conference, ISSCC. Digest of Technical Papers* (2007), pp. 520–521
72. C. Zhan, W.-H. Ki, A low dropout regulator for soc with high power supply rejection and low quiescent current," in *Proceedings of the 2009 12th International Symposium on Integrated Circuits, ISIC* (2009), pp. 37–40
73. C. Zhan, W.-H. Ki, An output-capacitor-free cascode low-dropout regulator with low quiescent current and high power supply rejection, in *IEEE Asia Pacific Conference on Circuits and Systems (APCCAS)* (2010), pp. 472–475
74. C. Zhan, W.-H. Ki, Analysis and design of output-capacitor-free low-dropout regulators with low quiescent current and high power supply rejection. IEEE Trans. Circuits Syst. I: Regul. Pap. **61**(2), 625–636 (2014)
75. E. Alon, J. Kim, S. Pamarti, K. Chang, M. Horowitz, Replica compensated linear regulators for supply-regulated phase-locked loops. IEEE J. Solid-State Circuits **41**(2), 413–424 (2006)
76. A. Amer, E. Sanchez-Sinencio, A 140 ma 90 nm cmos low drop-out regulator with -56db power supply rejection at 10mhz, in *IEEE Custom Integrated Circuits Conference (CICC)* (2010), pp. 1–4
77. M. El Nozahi, A. Amer, J. Torres, K. Entesari, E. Sanchez-Sinencio, High psr low drop-out regulator with feed-forward ripple cancellation technique. IEEE J. Solid-State Circuits **45**(3), 565–577 (2010)
78. A. Patel, G. Rincon-Mora, High power-supply-rejection (psr) current-mode low-dropout (ldo) regulator. IEEE Trans. Circuits Syst. II: Express Br. **57**(11), 868–873 (2010)
79. B. Yang, B. Drost, S. Rao, P. Hanumolu, A high-psr ldo using a feedforward supply-noise cancellation technique, in *IEEE Custom Integrated Circuits Conference (CICC)* (2011), pp. 1–4
80. E. Ho, P. Mok, Wide-loading-range fully integrated ldr with a power-supply ripple injection filter. IEEE Trans. Circuits Syst. II: Express Br. **59**(6), 356–360 (2012)
81. S. Ganta, C.-J. Park, D. Gitzel, R. Rivera, J. Silva-Martinez, An external capacitor-less low drop-out regulator with superior psr and fast transient response, in *IEEE 56th International Midwest Symposium on Circuits and Systems (MWSCAS)* (2013), pp. 137–140

82. Y. Liu, C. Zhan, W.-H. Ki, Fast-transient-response high-psr low-dropout regulator based on ultra-fast error amplifier and unity-gain buffer for portable applications, in *IEEE International Symposium on Circuits and Systems (ISCAS)* (2014), pp. 906–909
83. C.-J. Park, M. Onabajo, J. Silva-Martinez, External capacitor-less low drop-out regulator with 25 db superior power supply rejection in the 0.4–4 mhz range. IEEE J. Solid-State Circuits **49**(2), 486–501 (2014)
84. C.-J. Park, J. Silva-Martinez, M. Onabajo, Design techniques for external capacitor-less ldos with high psr over wide frequency range, in *IEEE 57th International Midwest Symposium on Circuits and Systems (MWSCAS)* (2014), pp. 342–345
85. Y. Lu, Y. Wang, Q. Pan, W.-H. Ki, C. Yue, A fully-integrated low-dropout regulator with full-spectrum power supply rejection. IEEE Trans. Circuits Syst. I: Regul. Pap. **62**(3), 707–716 (2015)
86. C.-H. Lee, K. McClellan, J. Choma, A supply-noise-insensitive cmos pll with a voltage regulator using dc-dc capacitive converter. IEEE J. Solid-State Circuits **36**(10), 1453–1463 (2001)
87. P. Hazucha, T. Karnik, B. Bloechel, C. Parsons, D. Finan, S. Borkar, Area-efficient linear regulator with ultra-fast load regulation. IEEE J. Solid-State Circuits **40**(4), 933–940 (2005)
88. J. Guo, K.N. Leung, A 6-μw chip-area-efficient output-capacitorless ldo in 90-nm cmos technology. IEEE J. Solid-State Circuits **45**(9), 1896–1905 (2010)
89. P.M. Furth, S. Krishnapurapu, S.H. Pakala, M.A. Haque, A 5.3 μa quiescent current fully-integrated low-dropout (ldo) regulator with transient recovery time enhancement, in *IEEE 56th International Midwest Symposium on Circuits and Systems (MWSCAS)* (2013), pp. 9–12
90. G.-Y.D. Harris, R. Ho, M. Horowitz, *The Fanout-of-4 Inverter Delay Metric* (1997)
91. D.G. Chinnery, Low power design automation. Ph.D. Dissertation, Electrical Engineering and Computer Sciences, University of California, Berkeley, 2006
92. D. Chinnery, K. Keutzer, *Overview of the Factors Affecting the Power Consumption*. (Springer, Boston, MA, 2007), pp. 11–53, http://dx.doi.org/10.1007/978-0-387-68953-1_2
93. P.R. Gray, *Analysis and Design of Analog Integrated Circuits*, 5th edn. (Wiley Publishing, 2009)
94. D.C. Daly, L.C. Fujino, K.C. Smith, Through the looking glass—the 2017 edition: trends in solid-state circuits from isscc. IEEE Solid-State Circuits Mag. **9**(1), 12–22 (2017)
95. A.J. Annema, B. Nauta, R. van Langevelde, H. Tuinhout, Analog circuits in ultra-deep-submicron cmos. IEEE J. Solid-State Circuits **40**(1), 132–143 (2005)
96. L.L. Lewyn, T. Ytterdal, C. Wulff, K. Martin, Analog circuit design in nanoscale cmos technologies. Proc. IEEE **97**(10), 1687–1714 (2009)
97. T. Ytterdal, Y. Cheng, T.A. Fjeldly, Wiley (2003), http://dx.doi.org/10.1002/0470863803.ch1
98. S.K. Saha, Compact mosfet modeling for process variability-aware vlsi circuit design. IEEE Access **2**, 104–115 (2014)
99. Z. Yan, P.I. Mak, R.P. Martins, Two stage operational amplifiers: power and area efficient frequency compensation for driving a wide range of capacitive load. IEEE Circuits Syst. Mag. **11**(1), 26–42 (2011)
100. G. Palumbo, S. Pennisi, *Feedback Amplifiers: Theory and Design*. (Springer, 2007), https://books.google.es/books?id=w4jqBwAAQBAJ
101. J. Miller, *Dependence of the input impedance of a three-electrode vacuum tube upon the load in the plate circuit*, ser. Scientific papers of the Bureau of Standards. Government Printing Office (1919), https://books.google.es/books?id=Q8CN1PJ4Zf0C
102. M. Yavari, O. Shoaei, F. Svelto, Hybrid cascode compensation for two-stage cmos operational amplifiers, in *IEEE International Symposium on Circuits and Systems*, vol. 2 (2005), pp. 1565–1568
103. W.-J. Huang, S.-I. Liu, Capacitor-free low dropout regulators using nested miller compensation with active resistor and 1-bit programmable capacitor array. IET Circuits Devices Syst. **2**(3), 306–316 (2008)
104. R. Eschauzier, J. Huijsing, *Frequency Compensation Techniques for Low-Power Operational Amplifiers*, ser. The Springer International Series in Engineering and Computer Science. (Springer, 2013), https://books.google.es/books?id=66zhBwAAQBAJ

105. R.G.H. Eschauzier, L.P.T. Kerklaan, J.H. Huijsing, A 100-mhz 100-db operational amplifier with multipath nested miller compensation structure. IEEE J. Solid-State Circuits **27**(12), 1709–1717 (1992)
106. J.H. Huijsing, R. Hogervorst, K.J. de Langen, Low-power low-voltage vlsi operational amplifier cells. IEEE Trans. Circuits Syst. I: Fundam. Theory Appl. **42**(11), 841–852 (1995)
107. C.C. Enz, F. Krummenacher, E.A. Vittoz, An analytical mos transistor model valid in all regions of operation and dedicated to low-voltage and low-current applications. Analog Integr. Circuits Signal Process. **8**(1), 83–114 (1995), https://doi.org/10.1007/BF01239381
108. E.A. Vittoz, *Pseudo-resistive networks and their applications to analog collective computation*. (Springer, Berlin, 1997), pp. 1131–1150, https://doi.org/10.1007/BFb0020305
109. J.E. Franca, Y. Tsividis (eds.), *Design of Analog-digital VLSI Circuits for Telecommunications and Signal Processing* (Prentice-Hall Inc, Upper Saddle River, NJ, USA, 1994)
110. X.L. Tan, K.C. Koay, S.S. Chong, P. Chan, A fvf ldo regulator with dual-summed miller frequency compensation for wide load capacitance range applications. IEEE Trans. Circuits Syst. I: Regul. Pap. **61**(5), 1304–1312 (2014)
111. D. Buss, Technology in the internet age, in *IEEE International Solid-State Circuits Conference, Digest of Technical Papers. ISSCC*, vol. 1 (2002), pp. 18–21
112. I. Akyildiz, W. Su, Y. Sankarasubramaniam, E. Cayirci, A survey on sensor networks. IEEE Commun. Mag. **40**(8), 102–114 (2002)
113. N. Ansari, T. Han, Freenet: spectrum and energy harvesting wireless networks. IEEE Netw. **30**(1), 66–71 (2016)
114. X. Wu, Y. Shi, S. Jeloka, K. Yang, I. Lee, Y. Lee, D. Sylvester, D. Blaauw, A 20-pw discontinuous switched-capacitor energy harvester for smart sensor applications. IEEE J. Solid-State Circuits **52**(4), 972–984 (2017)
115. R. Morello, S.C. Mukhopadhyay, Z. Liu, D. Slomovitz, S.R. Samantaray, Advances on sensing technologies for smart cities and power grids: a review. IEEE Sens. J. **17**(23), 7596–7610 (2017)
116. S. Jeong, Y. Chen, T. Jang, J.M.L. Tsai, D. Blaauw, H.S. Kim, D. Sylvester, Always-on 12-nw acoustic sensing and object recognition microsystem for unattended ground sensor nodes. IEEE J. Solid-State Circuits **PP**(99), 1–14 (2017)
117. M.H. Ghaed, G. Chen, R. u. Haque, M. Wieckowski, Y. Kim, G. Kim, Y. Lee, I. Lee, D. Fick, D. Kim, M. Seok, K. D. Wise, D. Blaauw, D. Sylvester, Circuits for a cubic-millimeter energy-autonomous wireless intraocular pressure monitor. IEEE Trans. Circuits Syst. I: Regul. Pap. **60**(12), 3152–3162 (2013)
118. P.P. Mercier, A.C. Lysaght, S. Bandyopadhyay, A.P. Chandrakasan, K.M. Stankovic, Energy extraction from the biologic battery in the inner ear. Nat. Biotechnol. **30**, 1240 EP (2012), http://dx.doi.org/10.1038/nbt.2394
119. S. Bandyopadhyay, P.P. Mercier, A.C. Lysaght, K.M. Stankovic, A.P. Chandrakasan, A 1.1 nw energy-harvesting system with 544 pw quiescent power for next-generation implants. IEEE J. Solid-State Circuits **49**(12), 2812–2824 (2014)
120. M. Yip, R. Jin, H.H. Nakajima, K.M. Stankovic, A.P. Chandrakasan, A fully-implantable cochlear implant soc with piezoelectric middle-ear sensor and arbitrary waveform neural stimulation. IEEE J. Solid-State Circuits **50**(1), 214–229 (2015)
121. E. Moon, D. Blaauw, J.D. Phillips, Subcutaneous photovoltaic infrared energy harvesting for bio-implantable devices. IEEE Trans. Electron Devices **64**(5), 2432–2437 (2017)
122. Y. Liao, M. Mollineaux, R. Hsu, R. Bartlett, A. Singla, A. Raja, R. Bajwa, R. Rajagopal, Snowfort: an open source wireless sensor network for data analytics in infrastructure and environmental monitoring. IEEE Sens. J. **14**(12), 4253–4263 (2014)
123. Y. Lee, D. Blaauw, D. Sylvester, Ultralow power circuit design for wireless sensor nodes for structural health monitoring. Proc. IEEE **104**(8), 1529–1546 (2016)
124. H. Wang, L. Dong, W. Wei, W.S. Zhao, K. Xu, G. Wang, The wsn monitoring system for large outdoor advertising boards based on zigbee and mems sensor. IEEE Sens. J. **PP**(99), 1–1 (2017)

125. I. Lee, Y. Lee, D. Sylvester, D. Blaauw, Battery voltage supervisors for miniature iot systems. IEEE J. Solid-State Circuits **51**(11), 2743–2756 (2016)
126. T. Jang, G. Kim, B. Kempke, M.B. Henry, N. Chiotellis, C. Pfeiffer, D. Kim, Y. Kim, Z. Foo, H. Kim, A. Grbic, D. Sylvester, H.S. Kim, D.D. Wentzloff, D. Blaauw, Circuit and system designs of ultra-low power sensor nodes with illustration in a miniaturized gnss logger for position tracking: part i—analog circuit techniques. IEEE Trans. Circuits Syst. I: Regul. Pap. **64**(9), 2237–2249 (2017)
127. T. Jang, G. Kim, B. Kempke, M.B. Henry, N. Chiotellis, C. Pfeiffer, D. Kim, Y. Kim, Z. Foo, H. Kim, A. Grbic, D. Sylvester, H.S. Kim, D.D. Wentzloff, D. Blaauw, Circuit and system designs of ultra-low power sensor nodes with illustration in a miniaturized gnss logger for position tracking: part ii—data communication, energy harvesting, power management, and digital circuits. IEEE Trans. Circuits Syst. I: Regul. Pap. **64**(9), 2250–2262 (2017)
128. Y. Lee, G. Chen, S. Hanson, D. Sylvester, D. Blaauw, Ultra-low power circuit techniques for a new class of sub-mm3 sensor nodes, in *IEEE Custom Integrated Circuits Conference* (2010), pp. 1–8
129. M. Alioto, Ultra-low power vlsi circuit design demystified and explained: a tutorial. IEEE Trans. Circuits Syst. I: Regul. Pap. **59**(1), 3–29 (2012)
130. Y. Ramadass, Powering the internet of things, in *IEEE Hot Chips 26 Symposium (HCS)* (2014), pp. 1–50
131. P.R. Panda, A. Shrivastava, B.V.N. Silpa, K. Gummidipudi, *Basic Low Power Digital Design*. (Springer, Boston, MA, 2010), pp. 11–39, https://doi.org/10.1007/978-1-4419-6388-8_2
132. T.C. Huang, C.Y. Hsieh, Y.Y. Yang, Y.H. Lee, Y.C. Kang, K.H. Chen, C.C. Huang, Y.H. Lin, M.W. Lee, A battery-free 217 nw static control power buck converter for wireless rf energy harvesting with $alpha$-calibrated dynamic on/off time and adaptive phase lead control. IEEE J. Solid-State Circuits **47**(4), 852–862 (2012)
133. S. Bandyopadhyay, A.P. Chandrakasan, Platform architecture for solar, thermal, and vibration energy combining with mppt and single inductor. IEEE J. Solid-State Circuits **47**(9), 2199–2215 (2012)
134. X. Liu, E. Sánchez-Sinencio, A highly efficient ultralow photovoltaic power harvesting system with mppt for internet of things smart nodes. IEEE Trans. Very Large Scale Integr. (VLSI) Syst. **23**(12), 3065–3075 (2015)
135. X. Liu, E. Sánchez-Sinencio, An 86 system with hysteresis regulation and time-domain mppt for iot smart nodes. IEEE J. Solid-State Circuits **50**(6), 1424–1437 (2015)
136. G. Chowdary, S. Chatterjee, A 300-nw sensitive, 50-na dc-dc converter for energy harvesting applications. IEEE Trans. Circuits Syst. I: Regul. Pap. **62**(11), 2674–2684 (2015)
137. G. Yu, K.W.R. Chew, Z.C. Sun, H. Tang, L. Siek, A 400 nw single-inductor dual-input-tri-output dc-dc buck-boost converter with maximum power point tracking for indoor photovoltaic energy harvesting. IEEE J. Solid-State Circuits **50**(11), 2758–2772 (2015)
138. J. Katic, S. Rodriguez, A. Rusu, A dual-output thermoelectric energy harvesting interface with 86.6% peak efficiency at 30 μw and total control power of 160 nw. IEEE J. Solid-State Circuits **51**(8), 1928–1937 (2016)
139. D. El-Damak, A.P. Chandrakasan, A 10 nw-1μw power management ic with integrated battery management and self-startup for energy harvesting applications. IEEE J. Solid-State Circuits **51**(4), 943–954 (2016)
140. P. Nadeau, D. El-Damak, D. Glettig, Y.L. Kong, S. Mo, C. Cleveland, L. Booth, N. Roxhed, R. Langer, A.P. Chandrakasan, G. Traverso, Prolonged energy harvesting for ingestible devices, Nat. Biomed. Eng. **1**, 0022 EP (2017), http://dx.doi.org/10.1038/s41551-016-0022
141. A. Paidimarri, A.P. Chandrakasan, A wide dynamic range buck converter with sub-nw quiescent power. IEEE J. Solid-State Circuits **52**(12), 3119–3131 (2017)
142. M.A. Abouzied, K. Ravichandran, E. Sánchez-Sinencio, A fully integrated reconfigurable self-startup rf energy-harvesting system with storage capability. IEEE J. Solid-State Circuits **52**(3), 704–719 (2017)
143. A. Das, Y. Gao, T.T.H. Kim, A 220-mv power-on-reset based self-starter with 2-nw quiescent power for thermoelectric energy harvesting systems. IEEE Trans. Circuits Syst. I: Regul. Pap. **64**(1), 217–226 (2017)

144. Q. Wan, Y.K. Teh, Y. Gao, P.K.T. Mok, Analysis and design of a thermoelectric energy harvesting system with reconfigurable array of thermoelectric generators for iot applications. IEEE Trans. Circuits Syst. I: Regul. Pap. **64**(9), 2346–2358 (2017)
145. L. Sumanen, M. Waltari, K. Halonen, A 10-bit 200-ms/s cmos parallel pipeline a/d converter. IEEE J. Solid-State Circuits **36**(7), 1048–1055 (2001)
146. J. Ramirez-Angulo, A. Lopez-Martin, R. Carvajal, A. Torralba, M. Jimenez, Simple class-ab voltage follower with slew rate and bandwidth enhancement and no extra static power or supply requirements. Electron. Lett. **42**(14), 784–785 (2006)
147. P.J. Hurst, S.H. Lewis, J.P. Keane, F. Aram, K.C. Dyer, Miller compensation using current buffers in fully differential cmos two-stage operational amplifiers. IEEE Trans. Circuits Syst. I: Regul. Pap. **51**(2), 275–285 (2004)
148. M. Yavari, O. Shoaei, A. Rodriguez-Vazquez, Systematic and optimal design of cmos two-stage opamps with hybrid cascode compensation, in *Proceedings of the Design Automation Test in Europe Conference*, vol. 1 (2006), p. 6
149. H. Aminzadeh, W. Serdijn, Low-dropout regulators: hybrid-cascode compensation to improve stability in nano-scale cmos technologies, in *IEEE International Symposium of Circuits and Systems (ISCAS)* (2011), pp. 2293–2296
150. H. Aminzadeh, M. Danaie, W.A. Serdijn, Hybrid cascode feedforward compensation for nano-scale low-power ultra-area-efficient three-stage amplifiers. Microelectron. J. **44**(12), 1201–1207 (2013), http://www.sciencedirect.com/science/article/pii/S0026269213001742
151. G. Rincon-Mora, P. Allen, Optimized frequency-shaping circuit topologies for ldos. IEEE Trans. Circuits Syst. II: Analog Digit. Signal Process. **45**(6), 703–708 (1998)
152. G.R.-M. R.K. Dokania, Cancellation of load regulation in low drop-out regulators. Electron. Lett. **38**, pp. 1300–1302(2) (2002)
153. M. Pulkin, G. Rincon-Mora, Stable low dropout, low impedance driver for linear regulators, U.S. Patent No. 6,573,694 (2003), http://www.google.com/patents/US6573694
154. J. Ramirez-Angulo, S. Gupta, I. Padilla, R. Carvajal, A. Torralba, M. Jimenez, F. Munoz, Comparison of conventional and new flipped voltage structures with increased input/output signal swing and current sourcing/sinking capabilities, in *48th Midwest Symposium on Circuits and Systems*, vol. 2 (2005), pp. 1151–1154
155. T. Man, C. Leung, K. Leung, P. Mok, J. Chan, Single-transistor-control low-dropout regulator, U.S. Patent No. 7,285,942 (2007), http://www.google.com/patents/US7285942
156. T. Man, K.N. Leung, C.Y. Leung, P. Mok, M. Chan, Development of single-transistor-control ldo based on flipped voltage follower for soc. IEEE Trans. Circuits Syst. I: Regul. Pap. **55**(5), 1392–1401 (2008)
157. Y. Tian, P.K. Chan, Design of high-performance analog circuits using wideband g_m-enhanced MOS composite transistors. IEICE Trans. **93-C**(7), 1199–1208 (2010), http://search.ieice.org/bin/summary.php?id=e93-c_7_1199
158. S. Chong, P. Chan, A flipped voltage follower based low-dropout regulator with composite power transistor, in *13th International Symposium on Integrated Circuits (ISIC)* (2011), pp. 472–475
159. K. Koay, S. Chong, P. Chan, A fvf based output capacitorless ldo regulator with wide load capacitance range, in *IEEE International Symposium on Circuits and Systems (ISCAS)* (2013), pp. 1488–1491
160. H. Kwa, X. Ji, B. Zhang, P. Chan, Amplifier circuit, and system incorporating same, U.S. Patent No. 7,265,632 (2007), http://www.google.com/patents/US7265632
161. G. Blakiewicz, Output-capacitorless low-dropout regulator using a cascoded flipped voltage follower. IET Circuits, Devices Syst. **5**, pp. 418–423(5) (2011), http://digital-library.theiet.org/content/journals/10.1049/iet-cds.2010.0431
162. P.Y. Or, K.N. Leung, An output-capacitorless low-dropout regulator with direct voltage-spike detection. IEEE J. Solid-State Circuits **45**(2), 458–466 (2010)
163. C.-M. Chen, T.-W. Tsai, C.-C. Hung, Fast transient low-dropout voltage regulator with hybrid dynamic biasing technique for soc application, IEEE Trans. Very Large Scale Integr. (VLSI) Syst. **21**(9), 1742–1747 (2013)

164. J. Guo, M. Ho, K.Y. Kwong, K.N. Leung, Power-area-efficient transient-improved capacitor-free fvf-ldo with digital detecting technique. Electron. Lett. **51**(1), 94–96 (2015)
165. K.N. Leung, M. Ho, J. Guo, P.Y. Or, Development of energy-efficient fast-transient cmos low-dropout regulators for soc applications, in *IEEE International Symposium on Circuits and Systems (ISCAS)* (2011), pp. 305–308
166. K. Luria, J. Shor, M. Zelikson, A. Lyakhov, Dual-mode low-drop-out regulator/power gate with linear and on-off conduction for microprocessor core on-die supply voltages in 14 nm, IEEE J. Solid-State Circuits **PP**(99), 1–11 (2016)
167. K. Luria, J. Shor, M. Zelikson, A. Lyakhov, 8.7 dual-use low-drop-out regulator/power gate with linear and on-off conduction modes for microprocessor on-die supply voltages in 14nm, in *IEEE International of Solid- State Circuits Conference—(ISSCC)* (2015), pp. 1–3
168. M. Yuffe, E. Knoll, M. Mehalel, J. Shor, T. Kurts, A fully integrated multi-cpu, gpu and memory controller 32 nm processor, in *IEEE International of Solid-State Circuits Conference Digest of Technical Papers (ISSCC)* (2011), pp. 264–266
169. K. Bult, G.J.G.M. Geelen, The cmos gain-boosting technique. Analog Integr. Circuits Signal Process. **1**(2), 119–135 (1991), https://doi.org/10.1007/BF00161305
170. A. Lopez-Martin, S. Baswa, J. Ramirez-Angulo, R. Carvajal, Low-voltage super class ab cmos ota cells with very high slew rate and power efficiency. IEEE J. Solid-State Circuits **40**(5), 1068–1077 (2005)

Index

© Springer International Publishing AG, part of Springer Nature 2018

J. M. Hinojo et al., *Internally Compensated LDO Regulators for Modern System-on-Chip Design*, https://doi.org/10.1007/978-3-319-75411-6

GPSR Compliance
The European Union's (EU) General Product Safety Regulation (GPSR) is a set of rules that requires consumer products to be safe and our obligations to ensure this.

If you have any concerns about our products, you can contact us on

ProductSafety@springernature.com

In case Publisher is established outside the EU, the EU authorized representative is:

Springer Nature Customer Service Center GmbH
Europaplatz 3
69115 Heidelberg, Germany

www.ingramcontent.com/pod-product-compliance
Lightning Source LLC
Chambersburg PA
CBHW060742150926
51501CB00003B/72
* 9 7 8 3 3 1 9 7 5 4 1 0 9 *